Television Station Operations and Management

Television Station Operations and Management

Edited by

ROBERT L. HILLIARD

Focal Press
Boston London

Focal Press is an imprint of Butterworth Publishers.

Library of Congress Cataloging-in-Publication Data

Television station operations and management / edited by Robert L. Hilliard.
p. cm.
Includes bibliographies and index.
ISBN 0-240-80027-3
1. Television stations—Management.
2. Television programs—Production and direction.
I. Hilliard, Robert L., 1925– .
HE8700.4.T45 1989
384.55′453′068—dc19 88-22078

British Library Cataloguing in Publication Data

Television station operations and management.
1. Television services. Production management
I. Hilliard, Robert L. *1925–*
384.55′4′0685
ISBN 0-240-80027-3

Butterworth Publishers
80 Montvale Avenue
Stoneham, MA 02180

10 9 8 7 6 5 4 3 2 1

Printed in the United States of America

To John and Sarah, whose courage, commitment, and consistency have made this world a better place.

Contents

Contributors

Thomas W. Cooper, Ph.D., University of Toronto, is Professor of Mass Communication, Emerson College.

Elizabeth Czech-Beckerman, Ph.D., Ohio State University, is retired Associate Professor of Radio–Television–Motion Pictures and former Assistant Dean of Arts and Sciences, University of North Carolina, Chapel Hill.

William Hawes, Ph.D., University of Michigan, is Professor and former Chair, Department of Radio–Television, School of Communication, University of Houston.

Robert L. Hilliard, Ph.D., Columbia University, is Professor of Mass Communication and former Dean of Graduate Studies, Emerson College. His previous positions include Chief, Educational (Public) Broadcasting Branch, of the Federal Communications Commission and Chair, Federal Interagency Media Committee.

Jeffrey Lukowsky, Ph.D., City University of New York, is Media Communications Consultant, Digital Equipment Corporation, and former Associate Professor, Emerson College, and Assistant Professor of Educational Communications, University of Wisconsin, Madison.

Kathleen P. Mahoney, Ph.D., Indiana University, former Chair, Department of Radio–Television, Ithaca College, is a Communication Research Analyst for Frank N. Magid Associates.

William Mockbee, a veteran of 35 years in broadcasting, has held positions of Sales Manager and Director of Operations for large market television stations, principally with the Westinghouse/ABC affiliate in Boston.

Preface

Most students studying television broadcasting plan to go into production: producing, directing, editing, performing, and writing. They are understandably disconcerted when I quote television executive Ralph Baruch's statement at a recent conference of the International Radio and Television Society: "If you're preparing your students in production, you're preparing them for unemployment."

Baruch and other professionals in the field emphasize that young people who wish to go into broadcasting should learn all the facets of the industry. Prepare students better for the realities that exist in broadcasting, they say.

That is what this book proposes to do. More and more colleges not only are offering but emphasizing courses in television operations and management. They know that that is where the jobs are: in administration, sales, research, marketing, promotion, advertising, and programming. This book is designed to provide the practical day-to-day skills and understanding necessary to function successfully in a television station in all of these areas.

Knowing the goals and potential results of one's labors is imperative to being a successful operations manager. Therefore, this book also provides the necessary background in studio and control room operations and equipment, in the transmission process, and in writing, directing, and performing. In operations and management one must deal with these areas every day, and an operations manager not familiar with them is seriously handicapped.

What if you insist on going into production despite the warnings of the television executives? One's chances of finding a job improve immeasurably by broadening one's skills and options. Knowing operations and management, as well as the production functions of a station, makes for a more desirable and valuable employee. It offers the station flexibility. Even if no production jobs are available at the moment, showing some knowledge of one or more of the other areas may help a job-seeker gain entry into the field and provide valuable experience until a production position becomes available. It gets a foot in the door. Besides that, many have found an operational job attractive enough to make a career in it.

The management, operations, and production functions in television, then, are not mutually exclusive. For example, upon examining the Leo Burnett Television Program Report listings in Chapter 4, it is evident that the executive producer or producer of a show very frequently also is the writer. It is therefore necessary for these management executives to know not only about writing, but how to write; conversely, the writer who does not know executive production management is limiting his or her potentials. The sales manager who does not know enough about writing, directing, and performing to supervise the making of a commercial will not remain a manager very long. The director who does not know how to work effectively with a station's budget supervisors, promotion people, sales executives, and other operations managers is not going to be successful, no matter how creative an artist.

Because most students will be working with individual television stations rather than with networks, the emphasis in this book is on what goes on in a station, both the affiliate and the independent. However, where appropriate, networks and independent producers are included in background and problem-solving discussions.

The purpose of this book is to provide practical formulas not only for understanding the television station business, but for getting and holding a job. The authors apply that approach to the content of each chapter, dealing with the specified subject in its real-life context: what the problems, options, and solutions are in the typical television station. But capable, hands-on application does not function in a vacuum. While being candid about the "bottom line," we also present the ethical considerations that go into making decisions. Choosing the best solution requires knowing all of the options.

All of the chapter contributors know that "quick and dirty" solutions may be successful in the short run and are practiced on all levels in the television industry. But we also know that each individual working in television has a heavy personal as well as corporate responsibility. Millions of people in the world probably are influenced more in their thoughts, feelings, and even actions by television than by any other single factor. Therefore, in every chapter we consider the impact on people, as well as on the company's profit-and-loss statement, of what we do in television.

I am grateful to the contributors to this book, Tom Cooper, Bill Mockbee, Kathy Mahoney, Betty Czech-Beckerman, Jeffrey Lukowsky, and Bill Hawes, for their patience and commitment, as well as for their knowledge and skills. I appreciate, too, my new-found editors and friends at Focal Press, Karen Speerstra, Phil Sutherland, and Judith Mara Riotto, whose ideas and cooperation made this book possible in its present form. I salute the dedication of faculty colleagues throughout the country in their efforts to prepare their students to be responsible citizens as well as capable practitioners in their work in the communications media. Finally, I am grateful to my students, whose visions for the future and whose needs—not necessarily yet understood or known by them—made this book necessary and a pleasure to write.

Robert L. Hilliard
Cambridge, Massachusetts
September 1988

1

Principles and Issues

Robert L. Hilliard

From the evidence of surviving records, human beings have always been fascinated with improving the art of communication. Even the records themselves have improved because of innovations in communication. Most recently, two types of improvement now frequently taken for granted have had an enormous influence upon mass communication. These are (1) the management of professional media operations (entire systems or individual segments of communication) and (2) television, the instantaneous mass distribution of live sounds and simultaneous moving images.

When human beings began writing histories of their societies more than 5,000 years ago, signs of their presence were left behind: scratchings, symbols, and pictures. Spoken language and music proved to be complementary accompaniments to this visual communication. As much as any other juxtaposition, the matching of words with pictures gave rise to modern communication. Audio/video interplay necessitated effective implementation of techniques, that is, operational skills, and, concomitantly, the effective management of the communication process.

Imagine the anticipation that swept through Europe when Gutenberg's movable type made possible mechanical reproduction rather than hand transcription of news. As culture and civilization spread throughout the world, so too did replicated forms of communication. Mass reproduction and widespread distribution of messages called for increasing numbers of people and departments to coordinate larger corporate processes. Operational management of time, space, and teamwork became essential to communication.

Recent modern communication bridges, from telegraph to telecommunication satellite, provide tools for planetary citizens to "reach out and touch" virtually everyone. The technology that brought long-distance visual communication to this process is television. Probably no other communication operation has been as influential as the proper coordination of local, national, and global television broadcasting and programming. In one sense, management of television operations is responsible, as much as any other profession, for the nature of thinking and attitudes of people worldwide.

In sum, television and telecommunication operations management is responsible for far more than what appear to be tiny stations from a global perspective. It contributes to the large network of people responsible for the health of global entertainment, education, commerce, attitudes, and even, to some degree, of world harmony. It also is partially responsible for the quality of reproduction, content, personnel, and technology of a substantial portion of human communication.

A PHILOSOPHY OF OPERATIONS MANAGEMENT

Prior to World War II, television was a crawling infant. During the 1930s the few stations on the air were experimental. Programming was limited to the amount of time that operators wished to transmit signals. Television sets in private households were few. Advertising was almost nonexistent.

Consequently, it is not surprising that formal theories of television operation were not commonplace. Early television managers were drawn from radio, just as many early radio performers came from the stage. The television operations director learned primarily through on-the-job rather than formal training.

Television management borrowed most of its attitudes from the traditions of radio management and corporate entrepreneurship. It is characterized by aggressiveness, assertiveness, ambition, and assiduousness. The prototype is further embodied by the image of a taskmaster whose concern was for the end product and its sales to larger and larger audiences. Rivalry for customers, which soon became rivalry for ratings, forced a philosophy that stressed achievement, growth, and competitiveness.

Over the years other skills have become important, including personnel psychology, research analysis, and image promotion. Needless to say, the primary skills that produce high-powered achievement, growth, and competitive success are still extremely significant.

The very nature of the television medium contributed to the development of particular types of operational skills. The simultaneous use of multiple cameras and microphones called for paradoxically different tendencies: on the one hand, a type of democratic teamwork among technicians and performers became essential; on the other hand, a more rigid, almost dictatorial relationship within the production/technician hierarchy took command. The split-second decision-making and deadlines of a live medium, pressured by costly studio time and rapidly changing news stories, left little time for reflective, debated, or team-analyzed decision-making.

Hierarchical authoritative management dominated during production, and questions were asked later. Casual, semidemocratic discussion might occur before or after but not during production. The quasimilitary chain-of-command structure, complete with high-pressure control rooms and headsets, would be reflected in the overall style of the station: a fast-paced, sometimes arbitrary, executive chain responsive to the competitive pressures of advertising, audiences, time, and technology.

Recently, a number of factors have modified some operational styles. While the basic goals of stations and networks remain intact—corporate profits, competitive superiority, perceived visibility as front-runner (in quality programming, news, and community service)—different techniques and attitudes have infiltrated many systems. More business school graduates, scientific researchers, and university products have improved quantitative research and compelled stations to rely more heavily upon demographic statistics and ratings. The marriage of television advertising agencies and other major businesses has given TV a greater awareness of management trends, whether written in books such as *The Peter Principle* or *In Search of Excellence,* or transferred by word of mouth.

The social status of women, minorities, strong unions, populist philosophies (win/win, democratic negotiation), and reactions to impersonal authoritarianism challenged the attitudes and traditions that dominated television operations. Moreover, unique lifestyles, variant self-disciplines, and local role models have varied from network to small town and from region to region. No one homogeneous style has dominated and many exceptions have proven the rule. Nevertheless, the early management philosophy—above all, success—remains the primary goal.*

Before management, in the abstract, or the individual operations director, in the particular, can deal effectively with the actual operational aspects of a television station, he or she must understand the broader issues affecting that station. The issues vary from new technological developments that may prove to be allies or enemies of the broadcast station, to the election of a United States president whose regulatory or deregulatory philosophy may force a change in staff development, sales approaches, and programming, among other things.

Even before these issues may be incorporated into plans and practices, there must be a philosophy or set of principles of television broadcasting under which they appropriately fit. Most evaluators of the media today regard television (and radio) as the most powerful forces in the world for affecting the minds and emotions of humankind. In many situations such impact has even controlled people's actions, such as purchasing a particular toothpaste or voting for a particular candidate. What should the role of broadcasting be? What should the role of the television station be? What should the role of the station one works for be?

Should television reflect public opinion or mold it? Should television take a leadership role or be passive, catering to and becoming part of the lowest common denominator? Should television control a country's political processes by controlling its information? Should television lead the way for social reform? Is there an unavoidable dichotomy between social responsibility and the profit-and-loss bottom line?

Answers to these questions must come from the backgrounds and philosophies of those who manage and operate television. Because the United States is one of the few countries in the world that does not own or operate any domestic television stations (although virtually all countries, including the U.S., operate overseas informational or, if you will, propaganda stations), the government does not dictate what management philosophy should be. To the degree that the government does decide on whether regulation of stations, as specified under the Communications Act of 1934, As Amended, and interpreted in the voluminous *Rules and Regulations of the Federal Communications Commission,* should be strong or weak, the government provides each station with greater or lesser options in determining its philosophical directions.

**The preceding material in this chapter was written by Thomas W. Cooper.*

REGULATION-DEREGULATION

Beginning in the early 1970s, under its Republican chair, Richard Wiley, the Federal Communications Commission (FCC) began what it called reregulation, examining those areas of its rules and regulations that seemed to put an unnecessary bureaucratic or reporting burden on television licensees. While some adherents to the public interest expressed concern that any lessening of regulatory requirements might open a floodgate leading to irresponsibility or, at least, a deterioration of station service in the public interest, most of the reregulations related to technical and minor matters that had minimal impact on the overall responsibilities of the stations. In the late 1970s, under its Democratic chair, Charles Ferris, the FCC began a transition from reregulation to deregulation, lifting additional numbers of requirements, but perhaps more importantly, internally eliminating some offices that had originally been established to strengthen the Commission's activities in the public interest, convenience, and necessity. One example was the abolition of the Office of Network Study, which for years had stood as a watchdog over station-network relations (the FCC does not license networks, but because it licenses stations, it can establish rules protecting stations from entering into agreements or actions with networks that might tend to give the networks excessive power). It abolished, as well, its Educational (Public) Broadcasting Branch, which had been set up almost 20 years earlier by Chairman Newton Minow to facilitate and oversee the development of noncommercial broadcasting, as one solution for Minow's concern with commercial television's "vast wasteland."

Under a new Republican president, Ronald Reagan, the Commission in the early 1980s embarked on a committed and consistent policy of strong deregulation. Its chair, Mark Fowler, took the initiative in leading the FCC to dismantle a number of regulations that had developed during a period of strong public concern: the mid-1960s through the mid-1970s. From mid-1981 to mid-1987, the FCC eliminated the requirement that broadcasters and cable operators file annual financial reports; simplified broadcast station application and transfer of ownership forms; authorized noncommercial broadcasters to air paid, promotional announcements for nonprofit groups and to expand the content of underwriting announcements; eliminated its antitrafficking rule (enabling speculators to buy and sell stations for the purpose of quick profit-making and not necessarily for the purpose of providing a responsible program service to the public); expanded the license period from three to five years; authorized multichannel multipoint distribution services at the expense of channels that had been reserved for nonprofit education use in the Instructional Television Fixed Service band; changed the identification (ID) rules to permit a station to identify itself with a community or communities in addition to the city of license; exempted cable systems from local rate regulation of tiered services; exempted broadcasters from the equal time rule when they sponsored their own political debates; virtually eliminated all regulations on call signs; eliminated the regional concentration rule which prohibited ownership of three stations when the primary service contour of any two within 100 miles of a third overlapped; pulled back its policy statement on children's programs (it never was enacted into a formal rule), enabling broadcasters to do virtually whatever they wished, including program-length commercials; eliminated the 7-7-7 restrictions on multiple ownership, permitting any given owner twelve television (TV), twelve frequency modulation (FM) and twelve amplitude modulation (AM) stations up to a national 25% service coverage; cut back on its previously careful judgment of the character, background, and financial qualifications of license applicants; eliminated the technical quality program performance standards for cable companies; modified the program-reporting requirements of stations; eliminated the requirement for station ascertainment of community needs; eliminated the requirements for a specified amount of nonentertainment (i.e., news and public affairs) programming; did away with its commercial time limitations; reduced its renewal requirements, which had included detailed information showing that a station had served the public interest during its license period, to a simple statement on a renewal form the size of a postcard; and eliminated much of its affirmative action policies, including preference to minorities and women in the distress sale of stations. The preceding is only a partial list of deregulatory actions in the 1980s relating to television.

Even after Fowler left as FCC chair, the marketplace concept that he had so effectively implemented continued. When a federal court remanded a Fairness Doctrine case to the Commission on the grounds that since the Fairness Doctrine had not been codified (enacted into law by Congress), the Commission was not obligated to enforce it, the Commission contemplated eliminating it.

Congress, strongly supporting a Fairness Doctrine, did pass such a bill. After it was vetoed by the President, and the Senate did not have the two-thirds vote to override, although that number was easily obtainable in the House, the FCC eliminated the Fairness Doctrine.

Richard Wiley, whose reregulation ostensibly triggered the continued process, noted in 1988 that he was somewhat dismayed to see how far it had all come. He termed the current process unregulation and wondered whether antiregulation would be the next step. For management, it is important not to forget that the philosophy of the party in power and of the president determine the regulatory stance of all agencies, and that the FCC could readily change from administration to administration. It is believed by some observers that it would take considerably more than the eight years Reagan had for deregulation to bring back regulation to where it was in the late 1970s.

For many years television stations operated under a formal code of self-regulation as well as informal individual self-regulation. The Code of Good Practices of the National Association of Broadcasters (NAB), to which most of its member stations subscribed, was challenged as unconstitutional and was abolished by an agreement between the Department of Justice and the NAB in 1985.

Voluntary self-regulation continues, although without too much consistency in most instances. For example, although the NAB Code frowned on unfair competitor advertising, one does not have to look too hard now to find it in television commercials. On the other hand, the self-regulatory ban on the advertising of hard liquor still holds strong. There has never been either a congressional or FCC prohibition on the advertising of hard liquor, but one rarely sees such ads on television, principally because stations are afraid that if they are carried, Congress might well pass a law prohibiting them and, possibly, other alcohol or alcohol-related product ads, thus creating imposed regulation tougher than self-regulation.

REGULATORY AND POLICY ISSUES

The continuing flux in regulation, based on its relationship to the political and philosophical climate, may make any given issue or requirement moot from one month or one year to the next. The following discussion of issues reflects some of the key areas that have come to the fore in recent years and that are likely to continue to be concerns of the television industry.

Fairness

Developing out of a relatively long history of court cases (i.e., the Mayflower and Red Lion decisions) and FCC policy statements (i.e., the "Blue Book"), the Fairness Doctrine simply held that if a station carried only one side of an issue that is controversial in its community of service, and if there are complaints about the coverage, and if the FCC investigates and finds that the complaints are valid, then the FCC may require the station to carry materials over a period of time that substantially present the other side or sides of the issue. Fairness has been one of the key battlegrounds between stations and consumer public interest advocates. Although eliminated in 1987, there is some likelihood that the doctrine may be restored by Congress or a new FCC.

Equal Time

Section 315 of the Communications Act of 1934 provides the basis for equal time for *bona fide* political candidates for a given office, with the exception of news and public affairs programs and debates sponsored by the station itself. In practice, the exceptions make the rule virtually inoperative, giving the station the option of promoting its own candidate(s) and leaving out others from coverage. The rule principally is effective in relation to advertising, where a station must provide equal advertising time (if it is available) at special rates for all candidates who have the money to buy such time.

Ascertainment

Praised by consumer-oriented groups during its years of application, the FCC Ascertainment of Community Needs requirement resulted in stations determining from appropriate members of the community the issues of importance in that community and showing the FCC what programming it did to make those issues visible. Although abolished under deregulation, ascertainment is one of the items that is likely to be revived in some form by a proregulatory administration and Commission.

Must Carry

With the abolition of the FCC's "must carry" rule by the courts and the further elimination of cable regulation by the Cable Communications Act of 1984, television stations are concerned that cable systems are no longer required to carry local stations (stations within a 50 mile radius of the city of cable service). Broadcasters fear that the elimination of a given station from a cable system that may well be serving some 50% and more of the homes in the area will seriously reduce its advertisers and possibly result in bankruptcy. One solution is the A-B switch, which can be installed on cable-wired sets to permit the viewer to easily switch the set to an outdoor antenna in order to pick up any TV broadcast station that the cable system may have eliminated.

Obscenity

A continuing area of concern and the subject of court cases (most particularly the Supreme Court decision in the WBAI "Seven Dirty Words" case in 1978) is obscenity, including sexual representation, on television. Although there is a profanity clause in the Communications Act of 1934, the FCC has had little success in obtaining a clear definition of obscenity. In 1988 the FCC established an obscenity rule that gave the Commission the prerogative to judge what is obscene based on its interpretation of national community standards of propriety, and ruled out what might be considered any indecent materials between the hours of 6:00 A.M. and midnight. Later that year Congress eliminated the early morning "window." This is the kind of regulation about which the federal courts, most likely the Supreme Court, will have the final say.

Children's Programming

In the so-called "kidvid" area, as in obscenity and other areas, Congress frequently not only expresses a public interest but threatens to or in fact does preempt the FCC. The pressures from citizen groups, the failure of the FCC to enact any rule or regulation (although in 1974 it did issue a Policy Statement), and the continuing interest of Congress suggests that some formal action on children's television might well be taken through legislative channels. Of special concern to stations whose principal orientation is the bottom line are the complaints about program-length commercials, host-selling, amount of advertising time, and allegedly harmful products promoted on children's programs. In 1988 President Reagan pocket-vetoed a bill that limited commercial time on children's television shows.

High-Definition Television

Most observers agree that the FCC was shortsighted in attempting to protect capital investments in television after World War II by preserving the National Television Standards Committee (NTSC) 525-line transmission standard. In the late 1980s attempts were made by the industry and the FCC to integrate the high-definition television (HDTV) technology into the U.S. The 1,125-

line system (some variance occurs with different systems) would bring the U.S. technical quality of television on a par with or better than the other systems throughout the world.

Satellites

A boon to television program distribution, one potential use of satellites may seriously affect the entire structure of television operation: direct broadcast satellite (DBS). Ostensibly, DBS could mean the direct distribution of signals from the network to the home, bypassing the local station entirely. In one sense, this would be similar to what cable is doing, or can do, picking up distant signals and relaying them directly to the home. Because local stations do have a following by virtue of their local orientation in news, commercials, and some feature and public affairs programs, they are carried by cable systems despite the lack of a must-carry rule. Technology, however, has at least in theory made the local station potentially outmoded.

Prime Time Access Rule

The Prime Time Access Rule (PTAR) has provided independent producers with outlets for their products, and provided local stations with a choice of programming that permits more time for local commercials and, concomitantly, more revenue. PTAR prohibits stations from carrying more than three hours of network programming during the 7:00 P.M.–11:00 P.M. slot (with the exception of a half-hour network news program that follows an hour of local news). Elimination of this rule would strengthen network control and, in the opinion of some, lessen local station independence and revenue.

Mandatory License

While the Cable Act of 1984 removes most restrictions from cable operation, the Copyright Act still gives the local station copyright control over any of its copyrighted materials that the cable system may include when it carries that station. Before the mandatory license, under which a cable system pays a standard fee, based on its income, to the Copyright Tribunal, which then distributes that income to stations, each station negotiated with each cable system for copyrighted material used. The mandatory license makes operational arrangements much easier for both cable and broadcasters. Lack of such a license puts individual TV stations in a stronger bargaining position.

Cross-Ownership

Based on the concept of diversity of information and ideas in a pluralistic political society, and justified by the argument of scarcity of channels, at least in quality and differentiation of content if not in quantity, the cross-ownership rule was established by the FCC to prohibit common ownership of a daily newspaper and a broadcast station in the same community. It is an offshoot of the multiple-ownership rule, noted earlier as 12-12-12. Prohibition of cross-ownership was unanimously upheld by the Supreme Court in 1978. The most dramatic challenge in recent years came in the late 1980s from Rupert Murdoch, who cross-owned newspapers and television stations in two different cities, New York and Boston. When Congress passed a rule that in effect prohibited the FCC from deregulating the cross-ownership rule, Murdoch successfully petitioned the courts to declare the rule unconstitutional. Its impact on operations is as significant as the multiple-ownership rule, which limits the number of stations one owner can have nationally, and the duopoly rule, which prohibits the ownership of two same-service or radio–TV stations in the same community. On the one hand, wealthy stations find that the rule inhibits their growth in a given market; on the other hand, other stations find that it protects them from quasimonopolistic competition.

Low-Power Television

The FCC's opening the spectrum to low-power TV (LPTV) stations that do not create interference with existing stations means the addition of perhaps 1,000 more stations in communities throughout the country. On the one hand, management and operational job opportunities and the development of new appropriate organization and process practices result from LPTV; on the other hand, existing local stations, while not necessarily concerned about competition for regional advertising by these limited-power stations, may well be concerned about competition for strictly local advertising.

Syndex

Broadcasters protect local station programming through syndicated exclusivity rules; that is, under syndex a given program may be contracted to only one station in the same service area. The FCC deregulated that requirement in 1980. However, the lack of such a requirement on cable created a situation where cable viewers may find the same program on a distant import signal as on a local station, thus creating competition to that station that conceivably could reduce its viewership and consequently its advertisers. In 1988 the FCC reinstated that rule, applying it to cable as well as broadcasting.

Cable

As noted or implied in several issues discussed above, cable has already affected television stations and might well affect them to a much greater degree. With as much as 75% of the country expected to be wired for cable by the mid-1990s, not only are many television stations worried about distant signal competition, but they are concerned that cable might buy off the principal TV entertainment attractions (as television did to radio in the early 1950s), forcing local stations as we now know them to either go off the air or to completely reorient their philosophies and formats.

Despite the continuing flux in national policy and regulation, some basic requirements remain in place and are likely to continue in one form or another, no matter the philosophy of the FCC. The licensing and renewal processes are probably the most important of these. The purpose of the Radio Act of 1927, which established the Federal Radio Commission, was to correct the chaos on the air by giving a government body the authority to license stations and establish frequency and power requirements, among other things. Stations had gone on the air with a pro forma license and without any federal controls over what frequency or power they used, thus creating overlapping signals and interference throughout much of the country.

To operate a station today, an applicant must first apply for a construction permit. The applicant must satisfy the FCC in three areas: legal, showing that the controlling interests, including members of the board, comply with the legal requirements of ownership; economic, showing that the applicant has enough money available to construct the station and operate it for a specified period of time; technical, showing that the transmission site, frequency, power, and antenna height will not cause interference to another station and that the equipment is type accepted (on a list issued by the FCC), enabling the station to operate without likelihood of equipment failure. In addition, the FCC asks for a statement of proposed programming and judges whether that programming will serve the interests of the viewers in the proposed coverage area. The extent to which these requirements are applied depends on whether the FCC is in a regulatory or deregulatory period.

Following construction of the station and approval by the FCC engineering staff of the technical aspects (during an *equipment test* period), the station is given approval for *program testing*, which is tantamount to being licensed. The actual license follows at a later date when the required paperwork at the FCC is completed.

After five years the television station license must be renewed. The FCC specifies the time for renewal applications, public notifications, and the submission of any additional required documentation. Depending on the regulatory mood of the FCC at the time, the station may have its license renewed by simply sending a postcard-sized form (as under deregulation continuing into 1989) stating that it has operated satisfactorily during the preceding license period, or the station may be required (as in the pre-1981 regulatory period) to present voluminous documents showing the extent to which it has operated in the public interest, and it also may have to counter any allegations to the contrary filed by the public with the FCC.

The student—and practitioner—is encouraged to consult *Broadcasting* magazine every week, plus other sources (see *Bibliography* at the end of this chapter), for comprehensive updates on the continually changing key issues and regulations.

CONTROLS

In addition to federal regulatory and policy offices, the television station has to consider two other major sources that exercise varying degrees of control: the sponsor and the public.

The Sponsor

In the early days of commercial television, specifically the 1950 decade of growth after the establishment of national network hookups prompted a rush of national sponsors from radio to the new visual medium, advertisers and advertising agencies controlled programming. With individual sponsors supporting an entire show, that sponsor—or on behalf of the sponsor, the ad agency—had the final word on all aspects of a program, even to the point of censoring individual lines as well as theme and subject. With the increasing costs of television production over the years and the concomitant increase in commercial time costs, individual advertisers began to pull back from program sponsorship and moved to shared sponsorship (i.e., every other week), multiple sponsorship (i.e., several sponsors for a given program), and in the 1980s, in many instances, to individual spots only. Except for relatively few remaining program sponsor situations, no single advertiser had effective control over a program, and the determination of content became the prerogative, principally, of the networks and, to a degree, the independent producers. Of course, locally produced programs were under the aegis of the local station.

That does not mean, however, that the sponsors became powerless. As Erik Barnouw points out in his book *The Sponsor,* the direct control of the advertiser from the "sponsor's booth" was no longer necessary, and, as a whole, sponsors continue to influence all phases of programming and continue to have a disproportionate impact on "American culture, mores, politics, and institutions." Sponsors continue to decide the life and death of a program on whatever ground they see fit, in addition to ratings. For example, when actor Ed Asner was attacked by the political and religious right for his support of medical aid to El Salvador rebels fighting against a dictatorial government in the early 1980s, some principal sponsors of his "Lou Grant" show withdrew, resulting in a cancellation of the series by the network. It was the capitulation of the networks to sponsors as well as to right wing political pressure that resulted in the infamous blacklist period of the 1950s—a period that the networks later abjectly apologized for, but apparently are still unable to put behind them in terms of principle and courage. Considering the greater financial and political strength of the networks, it is not surprising that the less affluent local station has little chance to withstand the pressures and controls of the sponsor.

The Consumer

A second significant area of control comes from citizens' groups. These groups run the gamut of the political, social, and economic spectra. The religious-political organization, the Moral Majority, has mounted a campaign more than once in objection to the theme or content or performer of a program. In the 1970s, following the court rulings that confirmed the right of the public to participate as intervenors in the FCC's licensing and renewal procedure, organizations representing Blacks, Hispanics, women, and other groups that systematically had been discriminated against in job, program, and outreach areas of television, successfully pushed many local stations to provide positions and programming for those constituencies.

While some local stations were challenged formally, in most instances an agreement between the station and the citizen group or groups was reached, thus avoiding the expenditures of time and money on attorneys, and possibly FCC hearings and court cases.

In addition to ongoing special interest groups, ad hoc citizen or consumer committees frequently confront a station on a specific issue. For example, members of a particular church in a particular community may organize to protest what they regard as excessive obscenity on a station; another group in the same town may protest what they regard as restrictive censorship of words and actions at the same station. In many cities parents have protested the quality of children's programs, sometimes in association with the national Action for Children's Television organization. A change in a station's format (i.e., an increase in entertainment programming and a decrease in news and public affairs programming) may trigger the formation or protest of a consumer group. Violence in programs continues to prompt concern and pressure from citizen organizations.

Under deregulation the successes of consumer and citizen organizations, including those promoting equal opportunity in jobs and programming, have waned, although their efforts have not necessarily decreased.

More effective, perhaps, in exercising pressure controls over stations are the traditional groups within the community. By the very nature of their being in financial, political, and social positions to operate a station, station owners are with few exceptions part of the elite and powerful in their city or town. It is understandable, then, that they might sincerely believe that the status quo—that is, the existing structure of the society in which they have achieved a position of recognition and success—is best for offering opportunity for growth for all people in the community. It is, therefore, no surprise that local stations are responsive to the conservative elements in their communities and run their operations and programming accordingly.

For example, in 1985, following a Boston station's cancellation of a program after pressure by a local religious group, a member of the Board of Directors of the local (New England) chapter of the National Academy of Television Arts and Sciences introduced the following resolution at a NATAS meeting:

> Freedom of the press, including television, is a cornerstone of American democracy. The United States is one of the few countries in the world where the media legally have freedom from the control of outside forces. To abandon that freedom under pressure from any group, public or private, no matter how laudatory its aims, is to subvert a basic principle of democracy and to undermine one of our cherished freedoms.
>
> As communicators in New England, we urge our broadcasting colleagues here and throughout the nation to stand firm, with courage and conviction, not to succumb, but to maintain the open marketplace of ideas that has marked freedom of speech, thought and press in our country. We pledge our support to our colleagues in this endeavor.

The members of the New England NATAS refused to even consider that resolution. One NATAS officer suggested that the resolution "was perceived by many as a criticism of the actions taken by member stations." Of course, it was! Not only, therefore, do outside pressures determine programming decisions, but the pressures of management tend to force individual broadcasters to give up the principle and courage to support freedom of ideas and speech.

Station operations are sensitive not only to all of the diverse pressures of a community, but management tends to integrate them with whatever philosophy guides the station.

Although, as indicated earlier, federal rules and regulations may change, stations should be fully acquainted with the Communications Act of 1934, with all of its amendments as they apply to television, and with the FCC Rules and Regulations—which are too voluminous to print or even summarize here. However, for an introductory overview, you may wish to consult in your library that part of the table of contents of Part 73 of the FCC Rules and Regulations that applies to TV stations.

BIBLIOGRAPHY

Most books are out of date virtually by the time they are printed. For continuing current information on issues, the following journals should be consulted on a regular basis:

Access, Better Radio and Television, Broadcast Management/Engineering (B.M./E.), Broadcasting, Cable Age, Cablevision, Channels, Communications and the Law, Community Antenna Television Journal (C.A.T.J.), Federal Communications Law Journal, Journal of Broadcasting, Journal of Communications, Media Report to Women, Satellite Communication, Telephony, Telecommunications Journal, TV/Radio Age, U.S. FCC Reports, and *Variety.*

These journals contain material of pertinence to all of the chapters and subject areas in this book.

Books that present basic issues, principles, and history include:

Bagdikian, Ben. *The Media Monopoly.* Boston: Beacon Press, 1982.

Barnouw, Erik. Three volume history of broadcasting in the United States. *A Tower in Babel* (to 1933), 1966. *The Golden Web* (1933–1953), 1968. *The Image Empire* (1953 to publication), 1970. New York: Oxford Press.

Head, Sidney W. and Christopher H. Sterling. *Broadcasting in America.* 5th ed. Boston: Houghton Mifflin, 1987.

Singleton, Loy. *Telecommunications in the Information Age.* 2nd ed. Cambridge, MA: Ballinger, 1986.

Sterling, Christopher H. and John M. Kittross. *Stay Tuned.* 2nd ed. Belmont, CA: Wadsworth, 1989.

Wells, Alan. *Mass Media and Society.* Lexington, MA: Lexington Books, 1987.

To keep up with the changing FCC Rules and Regulations, see *Broadcasting* magazine, *FCC Reports, The Federal Register* (a daily compilation of all federal agency actions), the *Code of Federal Regulations* (title 47 contains FCC actions), and *Pike and Fischer Radio Regulations,* a legal compilation of all FCC decisions. The safest, although hardly the least expensive, approach is to retain a communications law firm, which will keep track of all new rules and regulations in terms of their effect on your station.

2

Administration and Budget

Thomas W. Cooper and
William Mockbee

Modern television operation has called for a great degree of specialization. A manager's job may differ depending upon whether the facility is a large or small station, an independent or affiliated station, or a VHF or UHF station. The proliferation of television networks, stations, cable companies, and independent alternatives have created differing types of management roles in subspecialized categories which will be described throughout the chapter. Thus, at the outset, it is important to understand the differences among television stations and their structures.

TYPES OF STATIONS

Owned and Operated

Owned and operated (O & O) stations are part of the corporate possessions of a major television network. In other words, the network owns a local station, staffs it, pays the employee payroll, and thus operates the station as does any other owner of a facility. The abbreviated title, *owned and operated,* means that the station is a functioning module within a larger machine-like organization. Owned and operated facilities serve many purposes:

1. O & Os give the network full presence in major markets (e.g., New York, Chicago, Los Angeles, Philadelphia, San Francisco, Washington, Detroit, etc.).
2. O & Os provide valuable training areas for candidates who may be considered top management or star material for the climb to national influence.
3. O & Os generate much local income for the network by being highly competitive in the local marketplaces they serve.
4. O & Os allow the network to have experimental areas for developing new writing and production skills, where ideas may be pretested for the industry. For example, a new type of camera might be tested by a news department locally before being used nationally. Another example would be a program that, if popular in one market, might be distributed to others.

Managers of O & O stations may encounter special types of challenges. For example, upon occasion the general manager of such a station may wish to broadcast a special program of local interest throughout that market. To do so, the station manager would have to reschedule network programming or *preempt* the regular network schedule. While the network does not encourage rescheduling or preempting of its programs, network officials know that such actions help enhance the local image of the station.

The preemption of a network for a local show forces the network to seek an alternate source station in the market in order to maintain good relations and visibility, and to provide advertisers an outlet to reach people within that market. Ironically, a local special program may run directly opposite the parent network's show on another station.

Thus, the owned and operated station may occasionally compete against the network that owns it. Consequently, management of owned and operated stations may have pressures in scheduling, conflicting loyalty obligations, and other unique challenges.

The Member Affiliates

With station ownership limited to twelve TV stations per owner, the national television network could not exist without exposure in other markets. The network *affiliate* plays a key role. Although an affiliate is not owned by a network, the affiliation (or association) provides mutual benefits to both the station and the network. A network affiliate may have relations to the network similar to those of an O & O; however, the connection legally and pragmatically is much weaker. As there are over 200 U.S. markets as defined by the rating services, and as ownership is limited to twelve stations, a network can only penetrate 6% (twelve of two hundred) of all markets with O & Os. Without the affiliate, millions of homes each night would not receive their favorite programs from ABC, CBS, and NBC.

Just as the network can contract with outside producers for programming, the network affiliate can preempt the network and broadcast a local program, movie, or sporting event to attempt to increase its own local audience and income. This freedom frequently may be considered a threat by the network. Thus, managers of network affiliates are in many ways caught in the middle. They must negotiate with higher level managers with whom they have strong ties, and, to some degree, implement new program directions and trends that reflect the ideas of the network. At the same time they must exercise the autonomy to make decisions about what programming will best serve their local interests and those of their community. In recent years, as the affiliated stations have taken a stronger position in the local marketplace, they have often done so at the expense of network program areas. Thus, the network and affiliate sometimes are at odds over which program will air or transfer to another station.

Some critics have suggested that local considerations, rather than the larger audience, should replace network considerations in licensing O & O stations. By permitting stations to concentrate on the community of interest, it is suggested, increased competition would better serve the public interest.

The Independent

Many stations that do not choose to affiliate with a network may obtain a mixture of programming from film distributors, one or more networks, local production houses, national production houses, or special programming packages from cable, religion, sports, weather, and other services. Some periodically change their network affiliations. Independent station management frequently focuses on survival, growth, and multiple competition, problems made all the more difficult by being alone as an unsupported station. Managers of independent stations have the challenges of defining the station image, finding an audience for that station, maintaining viewers and ratings, and competing favorably with other independents and with changing network programming of O & Os and affiliates.

On the other hand, the independent station does not have to adhere to as many higher corporate decisions and has flexibility for experimental and creative programming not usually available to O & O stations or affiliates.

The growth of cable TV has had great impact upon the role of independent television stations, some of which have expanded their markets considerably, such as so-called super stations, and others of which have a decreased audience due to the number of new channels available to cable subscribers.

VHF or UHF

When the FCC first adopted technical rules for television, it dealt solely with very high frequency (VHF) television. The FCC first allocated space on the frequency spectrum for thirteen television channels. These channels, numbers one through thirteen (most sets receive channels two through thirteen, channel one having been subsequently assigned for non-public government use), are called the VHF band. When it was determined that ultrahigh frequency (UHF) space was available and could accommodate both video and audio signals, the Commission felt the problem of a limited frequency spectrum had been solved: more channels could be tuned in on home viewing sets (see Figure 2–1).

Unfortunately, the solution to the limited spectrum problem was not simple. Many manufacturers, believing that the weaker UHF signals could not compete favorably with the established VHF stations, did not rush to manufacture television sets that would receive all channels. People had to purchase special converters, which were costly and cumbersome. Several steps were required to fine-tune such converters, which might not (1) stay finely tuned, (2) receive the signal clearly, (3) pick up a signal more than twenty miles away, and (4) receive fresh, high-quality programming. UHF languished.

In its determination to make UHF work, the FCC adopted a policy of "dropping in" stations, which allowed specific UHF stations to overlap with major market areas where strong VHF stations already existed. An all-channel law in 1963 required all television sets made for sale in the United States to have the capability of receiving both UHF and VHF channels. Today, many UHF stations compete successfully for audiences in major cities. Some mixed markets (VHF/UHF) have one or two VHF network affiliates and a UHF network affiliate. Many markets support more than one UHF station.

Increasingly, the differences between UHF and VHF are disappearing. Moreover, the advent of cable's multichannel potentials makes UHF and VHF more closely resemble each other when compared with the more specialized (all-movie, all-sports, all-weather, all-news, all-music) cable channels.

Large or Small

While market sizes range from huge to miniscule, the operational and management purposes of television stations remain similar:

Figure 2–1 The spectrum of very high frequency (VHF) and ultrahigh frequency (UHF) channels.

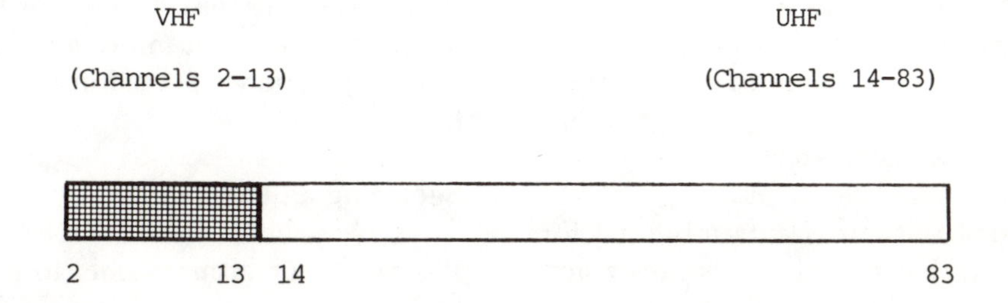

to maximize profits while minimizing cost. Although program schedules vary from one market to the other, the concept of the station, whether in a large or small market, is restricted only by the creative ability of management and its staff. Size of the station and size of the market do not determine operating skills, community service, nor all advertising presentations. There may be differences, however, in scale and emphases based upon the size of the market and the size of the station.

Students entering broadcasting should be aware that it is not only in the major market cities where success is possible. The real rewards in a broadcasting career lie within a station that allows for the personal expression of one's talents, gives a fair compensation for those talents, and offers the chance to grow within the organization.

Some job-seekers who have the luxury of choosing a station may take into account values and personal ideals: the type of programming, its quality, the honesty and professionalism of the management, the amount of freedom of expression, the business integrity of the station, and its leadership within the community.

Whatever the choice of station, students entering management should realize that management skills may be more easily obtained first within the smaller stations, particularly if there is room for advancement. The smaller stations usually offer greater opportunities to observe and assist management, and to obtain management training at an earlier stage of one's career. In smaller stations an individual often has a larger job scope and more flexibility to apply his or her skills. Larger stations frequently have larger degrees of specialization and limitations in job descriptions.

OWNERSHIP AND OWNERS

In the United States it is currently legal for one owner to have up to twelve television stations, twelve FM radio stations, and twelve AM radio stations simultaneously. These numbers, determined by the FCC, are continuing subjects of controversy. Some favor no limits. Others are concerned that ownership of television and radio may fall into the hands of a monopoly, duopoly, or small elite group of owners.

There are many stereotypes characterizing the image of media ownership. William Randolph Hearst, Rupert Murdock, and Lyndon Baines Johnson are among the many media owners to gain public attention. But owners cannot be stereotyped. The majority are responsible business people who are willing to incur financial risk by gambling for high stakes. They use the same devices as anyone else in business, borrowing, buying, and selling in order to operate. Owners must approve all sales and all modifications to the plant and equipment, and are ultimately responsible for the actions of their stations. They must hire and operate within the guidelines of Equal Employment Opportunity (EEO), affirmative action, Social Security, unemployment compensation, insurance, and, depending upon client reliability, delayed compensation.

An owner may operate just one small station, or, as indicated earlier, 36 broadcast stations. An owner may be on the job in one or more stations eighteen hours per day or may own many different enterprises and have nothing to do with the day-by-day operation of any given station. A local station owner may be the best local salesperson on the staff, work on the air, handle the bookkeeping, and serve as the station's engineer who climbs the tower to change a lightbulb. The FCC holds the owner responsible for the station, and whatever the owner's physical relationship to the station, it is wise for him or her to maintain accurate information on the station's operations, procedures, and bottom line.

Becoming an Owner

If a person or group wishes to purchase or build a station, there are several procedures that must be followed. First, potential licensees must file an application for a construction permit (CP) or transfer of ownership (when buying an existing station) with the FCC. Unless there are objections to the transfer or a filing for the same frequency by another bona fide applicant, the FCC usually approves. If there are problems, there may be hearings before the FCC prior to a final determination. The CP legitimizes the building of the station, the purchasing of equipment, construction of a tower, and setting up of facilities. Approval of a transfer authorizes the new owner to operate the station.

Financing the Purchase of the Station

In many ways, buying a station is a standard business transaction. Neither the FCC nor the government engages in funding (except for grants in certain instances for public television stations); instead, as with normal business transactions, most owners will apply to a bank for financing. Broadcast properties are considered to be low risk by banks because stations are tangible net assets that the bank may operate or sell.

It is still possible for a small entrepreneur to purchase one or more stations. However, with the advent of the superstation and the growth of giant market stations, TV properties are now bought and sold for billions of dollars. Not too many years ago, a television station could be purchased for less than $1 million.

STRUCTURE OF A TELEVISION STATION

Figure 2–2 shows the typical structure of a large television station. While the structure of stations vary and constantly change, positions such as general manager, sales manager, promotion manager, news director, operations manager, and so forth, tend to be universal in function, if not in title. These roles will be discussed in more detail later under *Administrative Staff*.

Three Factors That Influence Station Structure

Size

Size influences the specific details of a station's structure. Staff size depends upon the economic resources of the ownership, the number of stations in the marketplace, and the degree of commitment to programming over and above whatever may be supplied by network affiliation. In general, the smaller the market size, the smaller the station, and vice versa.

However, in many cases technology or automation has pruned down large staffs. The trend toward more computerized systems and high-tech development increasingly affects the number and type of staff chosen. In addition, the fractionalization of audiences in many large markets sometimes results in some stations (usually UHFs) surviving with relatively small staffs reaching relatively small numbers of viewers. Size, therefore, will not always mean market size, but may relate to the number of persons employed because of other factors. Working with a small staff is not necessarily a disadvantage. The smaller the staff, the better the opportunity for the employee who can perform more than one task to demonstrate talent and capability.

In a smaller market VHF or UHF station, it is not unusual to have a staff of thirty or less. In a major market UHF station, a staff of eighty or so is usual. For a major market VHF network affiliate, staff size may be as high as 350 to 400 employees. A major market UHF network affiliate may accrue up to 200. Staff size varies considerably and will accordingly determine the size and number of departments.

Power

VHF television stations operate within specific power guidelines set by the FCC. Tall antenna, level terrain, and compact populations permit greater coverage with lower power. Although VHF programs often look clearer and thus seem to be part of a more

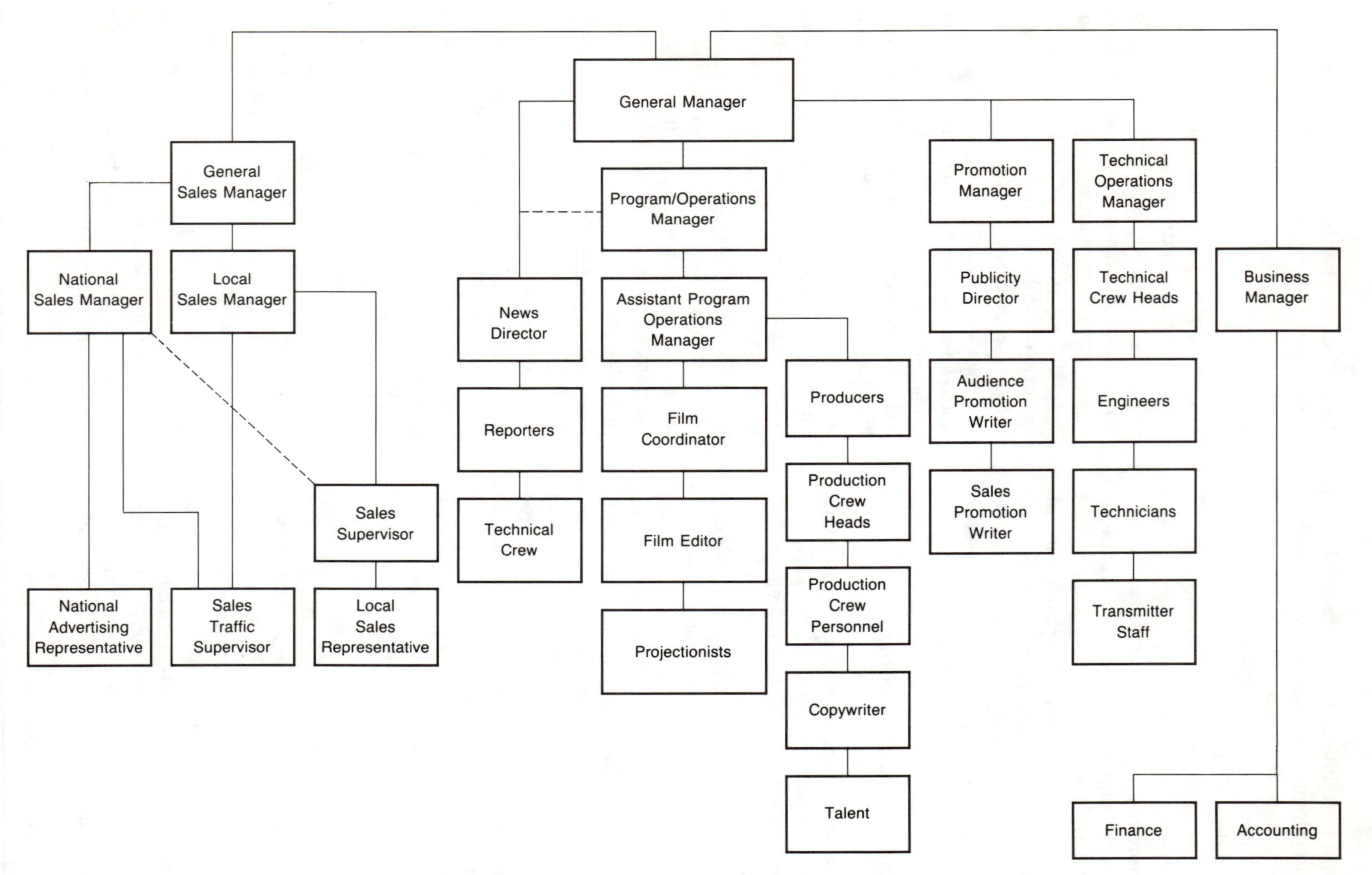

Figure 2–2 Structure of a typical large television station.

powerful signal, VHF stations do not require as much power as do UHF stations. The VHF part of the spectrum is wider and thus easier to fill with less power. Nor does VHF encounter as much interference.

UHF stations, however, operate within a much more limited part of the spectrum. They are subject to more interference and a larger drop-off in strength closer to the transmitting point. While channel 14 may follow channel 13 in the ordinal system of numbers, there is a wide gap in the broadcast spectrum between the two.

Through field strength test meters, scientific testing, and the continual ascertainment of signal strength by the FCC and private engineers, the television industry in North America enjoys a high degree of excellence in power control and quality. Fair use of the spectrum and ample coverage for almost all of the known needs result from these high standards.

Market

Over 3,100 counties, parishes, or their equivalents in the United States are ranked in terms of market size. Manufacturers of products once sold their goods to anyone located "in town." The advent of mass manufacturing, and subsequently mass marketing, carried sales considerations further than "around the corner and up the street." As mass media entered the selling picture, clear definitions of marketing areas were required. As the federal government grew and dispersed more services (and money) to the towns and cities, government departments also needed precisely defined locales. The U.S. Office of Management and Budget (OMB) ultimately divided marketing areas into Metropolitan Statistical Areas (MSAs), an arbitrary but useful system.

The two major U.S. rating services for television, Arbitron Ratings Company (ARB) and the A.C. Nielsen Company (NSI), using scientific methods and data provided by the OMB and others, have devised several key market definitions used by broadcasters. Where businesses draw customers only from the city of coverage, it is called a *home market*. Where the customers come from the metropolitan area surrounding the city, the market is called a *metro area*. Where the customers come from the suburbs as well as the metro and city areas, it is called a *designated market* (DMA) by Nielsen and an *area of dominant influence* (ADI) by Arbitron. Where the business is trying to reach a market that encompasses the entire reach of the station's signal, even in other cities, it is called the *total survey area* (TSA). In the latter case, the advertiser aims for 98% of the homes that can receive the station's signal.

The size, power, and market of a television station are closely interconnected and will influence its internal structure. But certain aspects of structure will transcend such categories. For example, all commercial television stations are necessarily hierarchical and compartmentalized. There is always need for one group of people to supervise programming and another group to ensure that the intervals between programs, the "slots" for commercials, are filled. No matter how small or large the station, there is some chain-of-command mechanism, frequently with middle managers who work closely with executive or general managers. This vertical structure has an accompanying structure and pay scale in which executives may receive ten or even one hundred times more financial reward for their services than entry level employees. Interns, of course, are frequently unpaid assistants who are learning the basic skills within the industry by observation and part-time assistance.

In addition to this vertical structure, there is a horizontal structure in which the basic function of a station, such as news, programming, and finances, are separated. This horizontal area is even further subdivided. For example, finances can be separated into business operations, sales, marketing, bookkeeping, etc. More detailed structure is provided later in this chapter.

THE BUDGET

In both commercial and noncommercial broadcasting, management of the budget is extremely important. Perhaps the most important skill for any manager to acquire is proper maintenance and "protection of the bottom line" (overall income and expenditure balance). Much to the surprise of many new employees, the budget determines policy.

A charted budget can be described as a floor plan that provides the foundation for designing a station's operation. For it to be complete, a budget requires input from each department. From this input management executives may obtain general estimates of how much *income* the station will generate and what *expenses* the station will incur (see Figure 2–3). The expenses incurred by each department include its operating costs. These may include all of the salaries and wage benefits paid to employees, unemployment compensation costs, cost-of-living agreements, purchase of new equipment, maintenance of present equipment, and hidden expenses (e.g., tower painting, new antenna needs, new transmitter, etc.).

The budget also includes subdivisions under income such as network affiliation payments, syndication agreements, the sale of products (programming) to other stations,

Figure 2–3 Typical annual income and expenditures, large market television station.

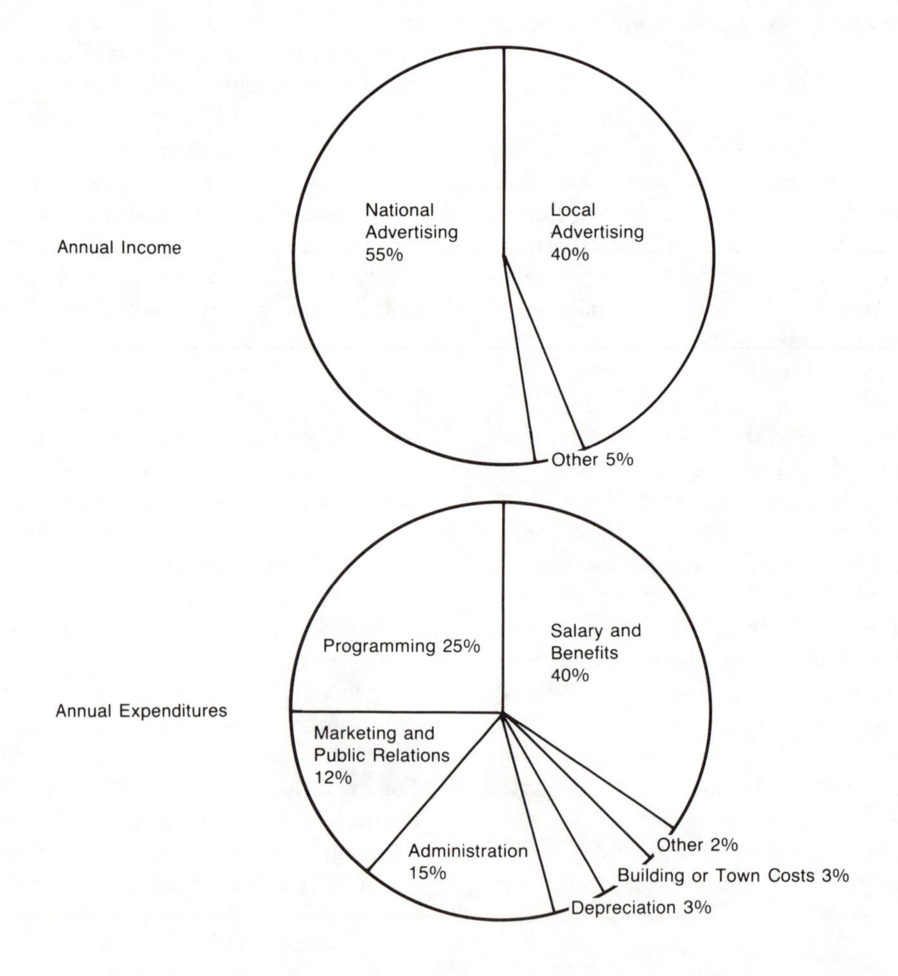

the sales generated by the national, regional, and local sales forces to advertisers, and many other lesser forms of income. A budget should also include a provision for additional income should the station be successful with ratings and thus able to charge more for each spot announcement. Moreover, unexpected expenses should be built into a strong budget so that deficit spending is not a disruptive surprise factor.

Obviously, the larger the station the greater the budget. It is not unusual for a small market station to have an annual budget equal to a major market station's weekly budget. In either case, all costs must be shown and specifically budgeted (see Figure 2–4). Every expense and income must be traced and projections made about anticipated revenue and expenditures. A mistake of a single percentage point may be costly to either the small or large operator. While the dollar amounts may be different, the risks are equally great.

No one chart can show how a budget will actually function, because there are always uncertain variables. Writing an accurate budget must allow for both unanticipated excess income and unexpected loss. For example, consider the following situation: if television station A's local news program is not number one in the marketplace ratings, that station will not draw "top dollar" (the best price) for advertising in that newscast. However, top rated station B suddenly loses a key anchor personality, the station's audience does not like the anchor's replacement, and some of that station's viewers switch to station A. When the new ratings are published, A's local news program is number one. A's advertising rates for that time period increase to reflect its new status. This is additional, unexpected income. Conversely, one or more programs might unexpectedly lose some income. Thus, while a manager budgets for a certain percentage of profit, that manager must prepare also for possible changes.

There are many factors, foreseen and unforeseen, that affect a station's income. Many successful stations do not profit every year. When a station loses money over the short term, new factors can result in recovery; business may turn around. Nevertheless, cost-cutting procedures are usually introduced in such situations, or changes in management and sales approaches may be effected.

Sometimes, however, market conditions can create loss situations that are far beyond a station's control: a mill may close that employs a substantial number of viewers in one's market; a new shopping center that is expected to attract large numbers of suburban customers may fail to draw major department stores and fall through as a project. Advertisers are careful observers of station audiences and choose discriminately among the many outlets for their advertising expenditures.

To anticipate, monitor, and sometimes prevent financial disaster, a cost control system is a necessity. In most stations computers are used to provide periodic (e.g., daily or monthly) printouts on income and expenses (see Figure 2–5). The allocation of funds is divided into budget or cost control centers. While each department may share the same end-code number for stationery suppliers, for example, the start-code number will be different. Thus, each department will have unique coding and can keep track of its bills. Moreover, the accounting department may observe at a glance how any given department is spending its share of the budget. A sample printout would show the following:

1. Department name
2. Code number for the department
3. Each budget allocation for that department
4. Amount budgeted for the month and year-to-date (YTD)
5. Expenditures for the month and year-to-date
6. Amount over or under budget for each allocation code
7. Running total for each divisional part of the department

WQTV, INC.
BALANCE SHEET--ASSETS — SERIAL NUMBER: 00033

DESCRIPTION	CURRENT YEAR	PREVIOUS YEAR
CURRENT ASSETS:		
CASH AND BANK	0	0
ACCOUNTS RECEIVABLE:	0	0
TRADE (NET)	0	0
OFFICERS AND OTHER RECEIVABLES	0	0
PROGRAM LICENSE (CURRENT)	0	0
PREPAID EXPENSES	0	0
TOTAL CURRENT ASSETS	0	0
PROPERTY AND EQUIPMENT:		
CONTROL ROOM & STUDIO EQUIPMENT (NET)	0	0
TRANSMITTER EQUIPMENT (NET)	0	0
TOWER, BASES 7 EASEMENTS (NET)	0	0
LEASEHOLD IMPROVEMENTS (NET)	0	0
OTHER PROPERTY AND EQUIPMENT (NET)	0	0
TOTAL PROPERTY AND EQUIPMENT	0	0
OTHER ASSETS:		
PROGRAM LICENSE - LONG-TERM (NET)	0	0
SECURITY DEPOSITS	0	0
TRADE RECEIVABLES (BARTER)	0	0
INTER-COMPANY ACCOUNTS RECEIVABLES	0	0
TOTAL OTHER ASSETS	0	0
INTANGIBLE ASSETS:		
F.C.C. TELEVISION LICENSE/GOODWILL (NET)	0	0
TOTAL INTANGIBLE ASSETS	0	0
TOTAL ASSETS	0	0

Figure 2–4 Sample station balance sheets displaying assets *(left)* and liabilities *(right)*. (Courtesy of WQTV, Channel 68, Boston. Actual figures remain confidential by request.)

WQTV, INC.
BALANCE SHEET--LIABILITIES &
STOCKHOLDERS' EQUITY SERIAL NUMBER: 00033

DESCRIPTION	CURRENT YEAR	PREVIOUS YEAR
CURRENT LIABILITIES:		
ACCRUED EXPENSES	0	0
AMOUNTS WITHHELD FROM EMPLOYEES	0	0
FEDERAL INCOME TAX PAYABLE	0	0
ACCOUNTS PAYABLE	0	0
ACCOUNTS PAYABLE - PROGRAM LICENSES NOW DUE	0	0
PROGRAM LICENSE PAYABLE (CURRENT)	0	0
NOTES PAYABLE (CURRENT)	0	0
DEFERRED TRADE INCOME	0	0
TOTAL CURRENT LIABILITIES	0	0
LONG-TERM LIABILITIES:		
NOTES PAYABLE (LONG-TERM)	0	0
PROGRAM LICENSE PAYABLE (LONG-TERM)	0	0
INTER-COMPANY PAYABLES	0	0
TOTAL LONG-TERM LIABILITIES	0	0
STOCKHOLDERS' EQUITY:		
CAPITAL STOCK	0	0
RETAINED EARNINGS	0	0
PAID-IN-CAPITAL	0	0
TOTAL STOCKHOLDERS' EQUITY	0	0
TOTAL LIABILITIES & STOCKHOLDERS' EQUITY	0	0

WQTV, INC.
ENGINEERING EXPENSES PROFIT & LOSS STATEMENT SERIAL NUMBER: 00027 DATE: 4/30/86

THIS STATEMENT REFLECTS THE HISTORICAL (03/31/85 AND BEFORE) INCOME AND EXPENSE FIGURES OF BOSTON STAR BROADCASTING FOR REPORTING PURPOSES ONLY.

CURRENT MONTH				YEAR TO DATE		
ACTUAL	HISTORY	VARIANCE	DESCRIPTION	ACTUAL	HISTORY	VARIANCE
0	0	0	SALARY	0	0	0
0	0	0	SALARY RELATED EXPENSE	0	0	0
0	0	0	COMPANY AUTO EXPENSE	0	0	0
0	0	0	CONSULTING EXPENSE	0	0	0
0	0	0	CORPORATE ENGINEERING EXP	0	0	0
0	0	0	DUES & SUBSCRIPTIONS	0	0	0
0	0	0	EQUIPMENT MAINT & REPAIR	0	0	0
0	0	0	MISC ENGINEERING EXPENSES	0	0	0
0	0	0	PARTS & SUPPLIES	0	0	0
0	0	0	TRADE EXPENSE	0	0	0
0	0	0	TRANSMITTER RENT EXPENSE	0	0	0
0	0	0	TRANSMITTER TUBES EXPENSE	0	0	0
0	0	0	TRANSMITTER UTILITIES EXP	0	0	0
0	0	0	TRAVEL & ENTERTAINMENT	0	0	0
0	0	0	TOTAL ENGINEERING	0	0	0

PROGRAMMING EXPENSES PROFIT & LOSS STATEMENT SERIAL NUMBER: 00027 DATE: 4/30/86

THIS STATEMENT REFLECTS THE HISTORICAL (03/31/85 AND BEFORE) INCOME AND EXPENSE FIGURES OF BOSTON STAR BROADCASTING FOR REPORTING PURPOSES ONLY.

CURRENT MONTH				YEAR TO DATE		
ACTUAL	HISTORY	VARIANCE	DESCRIPTION	ACTUAL	HISTORY	VARIANCE
0	0	0	SALARY	0	0	0
0	0	0	SALARY RELATED EXPENSE	0	0	0
0	0	0	COMPANY AUTO EXPENSE	0	0	0
0	0	0	CONSULTING EXPENSE	0	0	0
0	0	0	DELIVERY EXPENSE	0	0	0
0	0	0	DUES & SUBSCRIPTIONS	0	0	0
0	0	0	FILM AMORTIZATION EXP.	0	0	0
0	0	0	FILM PRINT	0	0	0
0	0	0	MISC. PROGRAMMING EXP.	0	0	0
0	0	0	MUSIC LICENSE FEES	0	0	0
0	0	0	PRODUCTION EXPENSE	0	0	0
0	0	0	TRADE EXPENSE	0	0	0
0	0	0	TRAVEL & ENTERTAINMENT	0	0	0
0	0	0	VIDEO TAPE	0	0	0
0	0	0	TOTAL PROGRAMMING	0	0	0

Figure 2–5 Sample monthly expenditure and expense printouts for an engineering division *(top)* and a programming division *(bottom).* (Courtesy of WQTV, Channel 68, Boston. Actual figures remain confidential by request.)

The central computing system is given this information by the accounting department. Together, they oversee the approved bills, pay out the monies to those vendors who have supplied services, and enter the data into the computer under proper coding. Smaller stations unable to afford a full-time computer may have a time-share computing arrangement. Although they may not have printout information as frequently as they desire, it is still possible to successfully monitor the overall flow of money within the division, department, and station.

Trial budgets can be fed into a computer. Similarly, market reviews permit assessment of the competition and their costs. Long- and short-term projections may be factored to enable clear-cut budget decisions to be made. In short, the art of balancing budgets is constantly made more scientific by computation technology, which allows an increase in the frequency of monitoring expenditures and an increasing variety of other computer applications.

Budget managers not only must understand or at least be able to supervise the generation of computer printouts, but must be able to handle the complex relationships with middle managers who lobby for larger portions of budget expenditures for their departments.

THE STATION MANAGER

Background

From the organizational chart (Figure 2–2) it is obvious that the most important role at a television station is that of the station manager or general manager. Although the use of the terms vary at different stations, they are generally thought of as interchangeable in the field as a whole. The station manager reports to the owner either directly or, in a large multiproperty organization, indirectly through another executive officer. All power and responsibility flow from the ownership to the manager. The manager in turn delegates the functions to various department heads who make up the structure of the station.

The manager must maintain knowledge of all station activities. Budgetary control and the flow of the monies are part of the manager's responsibility. The ultimate decision in hiring and firing is a major concern for station managers.

Many managers hold regular staff meetings with their employees to ascertain problems and obtain suggestions. These meetings also help the employees to better understand the operations and problems of the station. One of the manager's primary responsibilities is to maintain smooth professional relationships among other executives and middle managers and a constant internal communication flow among those who coordinate major activities.

A manager has direct control over and responsibility for the various departments that constitute the total station. These include:

1. Office and Administrative
2. Programming
3. News
4. Operations/Engineering
5. Sales
6. Research
7. Promotion
8. Legal
9. Accounting/Billing
10. Employee Relations

Not all stations are identical, of course, but most contain these departments or, in smaller stations, two or more of these departments under one head.

The station manager frequently comes from a strong sales background. Demonstrating a capacity for constructing and implementing realistic budgets, and increasing the revenue potential of a station, is a key to current successful station leadership. Acceptance in the community that the station serves is also an important criterion for selecting a manager. Some managers come directly from a programming background or from the ranks of research. Some are transplanted from radio broadcasting. Others come from industries not directly related to broadcasting, but where they have demonstrated organizational skills and solid maintenance/protection of the bottom line. Whatever the background, financial and organizational experience is essential. In addition, the manager must possess skills of leadership, ability to issue work orders for jobs to be done, and the patience and wisdom necessary to deal with issues and employees on an individual as well as a group basis. A strident voice does not earn respect. A lack of ability, which initially may be covered over by charismatic personality or style, usually does not survive too long at the management level.

Policy statements, budget planning, engineering needs, programming requirements, employee benefits, community demands, union relationships, policy implementation, and, finally, the overall direction of the station, must be set by the station manager, coordinated by the station manager, and led by the station manager. Such a person will fail to succeed if, as an overall manager of other managers, he or she does not hire competent middle-level and even lower-level managers for the station's various departments.

Good working relationships are necessary not only with these other managers, but with the ownership of the station. There is no one typical relationship between a manager and an owner. One type of ownership may simply ask to see the manager at the end of the year, closely inspect the bottom line, inquire about the station's ability to follow FCC guidelines, and decide that the manager may be granted full autonomy for another year. Other owners wish to become involved in the day-to-day operations of a station, from editorializing over the air to auditing the books. A manager must be flexible enough to accommodate different styles of ownership, meet different levels of expectation, and at the same time fairly represent employees to the owner and, conversely, the ownership to the employees.

Relationships to Employees

Hiring, Promotion, and Release of Employees

Much of a manager's time may be devoted to the turnover of personnel. Knowing when to choose and whom to choose is vital to the bloodstream of any organization. How long to retain, when to promote, and whether to lay off employees must, whatever one's personal preference, serve the goals of the larger organization. Both labor and management have developed policies to protect both managers and employees.

Unions have grown in importance. They negotiate contracts with management that may stipulate procedures and establish grounds for grievance in hiring, firing, laying off, and promotion. Evaluation processes, for example, include annual employee reviews in which the perspectives of colleagues and supervisors, as well as individual work production, are taken into account. More informal types of evaluation take place in the form of feedback from job, social, and community activities. A skillful manager must know how to interpret feedback regarding each employee and, to the extent it is possible, understand the employee personally—that person's goals, the personal factors that may influence job performance, background, and training, and potential to help the company.

For a period prior to the Reagan administration of the 1980s, management had been increasingly responsive to government legislation regarding equal opportunity and nondiscriminatory practices. Social and political pressures concerning affirmative action, nondiscrimination on the basis of age, race, or sex, and sensitivity to the handicapped began to have significant impact on hiring practices. Technically, management must still remain aware of government regulations regarding hiring practices.

Unions

A manager should fully understand the nature and effects of unionization. The first step is to dismiss stereotypes, such as the following: that one union controls all of television, that all unions are alike, that all jobs are uniformly unionized, that management always dominates unions, or that all unions ultimately dominate management. In fact, there are different unions that apply to different professions. In some parts of the country and in certain jobs, union membership is mandatory; in other situations membership may be voluntary. And in still others, there are nonunion shops.

Some groups to which professionals belong are merely organizations or associations without the legal strength of unionization. Some unions have become nominal or routinized in their functions. Others have been effective in implementing change, and maintain strong lobbies.

Any given station operations manager may deal with one or more unions. Perhaps the union best known to the public is the American Federation of Television and Radio Artists (AFTRA), which represents television performers. Although not involved with television in the medium's early days because TV was live, the Screen Actors Guild (SAG) became a representative for many performers when television dramatic production moved to Hollywood and became predominantly filmed or taped. Today many performers belong to both AFTRA and SAG. The American Guild of Variety Artists (AGVA) is the union for musical and dance and other variety performers. The Writers Guild of America (WGA) is the organization of TV and film writers, especially those preparing network and syndicated program scripts.

A number of unions represent television technical personnel. The oldest is the International Brotherhood of Electrical Workers (IBEW), formed over 60 years ago in the early days of radio. NABET was established by NBC engineers as the National Association of Broadcast Engineers and Technicians, and later changed the word *engineers* to *employees* to serve a large constituency. When television production moved largely to Hollywood and the film industry became involved, another union, the International Alliance of Theatrical Stage Employees and Motion Picture Machine Operators of the United States and Canada (IATSE), entered the television industry.

A number of nonbroadcast unions in the clerical, service, and management fields represent their members at a number of stations.

The manager must be alert to the differences, subtleties, and powers of unions and must maintain fair practices toward employees. It is difficult to say what impact the further growth of cable television and other restructuring of the television industry will have upon unionization, but it is almost certain that managers increasingly will have to master collective bargaining and negotiation.

Training Programs

Managers may develop, or delegate to others for development, programs that allow employees to learn on the job and outside the station (i.e., conferences, workshops, college courses) to increase their skills. Professionally, such training programs enhance both individual contributions and group teamwork. Psychologically, such programs may give employees a sense of being cared about by management, and greater motivation to do their best for the station.

Internal Communication

Ironically, many television managers agree that the internal communication among professional communicators is mediocre, at best. Individuals who communicate effectively to millions of people often spend hours unsuccessfully trying to communicate with each other within one station. A manager must be effective with internal as well as external communication. Much of the job requires a careful coordination of detail, both in policy and in practice: instant, clear contact with staff, and efficient use and trafficking of time, important memos, external correspondence, appointments, telephoning, and clerical assignments. Delegation is a key responsibility among managers, so that many assignments involve correspondence and the ability to speed-read reports, write succinct and penetrating memos, and respond immediately to written and telephone communications. Some managers develop efficient systems for precise handling of mail, memos, telephones, and intercoms. Others work more spontaneously and develop a rhythm that is subordinate to the larger rhythms of the overall organization. Clear, concise, and catalyzing communication is essential to the job.

Relationships to Middle Management

A station manager must be sensitive to the needs and perspectives of all those at other levels of management. The station manager must be careful with the perception by lower level managers of favoritism, nepotism, or prejudice. Personal and professional relationships with each divisional director, whether the chief engineer, the program director, sales director, or the news director, are reflected in the overall operation of the station; harmonious relations are important. Some station managers may watch designated departments more carefully than others. For example, if the manager's background is in sales, he or she may pay more attention to the specifics of how revenues are rising or falling. If news is the primary product of the local station, the manager may watch the news ratings and meet frequently with the news director about competitive strategies. Most general managers are in touch on a day-to-day basis with each of their managers and may have daily or weekly meetings with core staff. A station manager is only as strong as the immediate team or circle with whom that manager communicates. Without trust, respect and cooperation at the higher levels of organization, it is difficult to maintain fluidity throughout the other levels of organization.

Often leadership is demonstrated by a general manager to other managers by virtue of availability. When an important news story is breaking that may require continuous coverage, a station manager may stay in the newsroom simply to demonstrate responsibility and support for news personnel. Questions may arise: Should the station go back to local programming? Should the network coverage be honored? Should local announcements explain the late breaking story to the local audience? Should news employees work overtime if the story continues around the clock? A responsible station manager is highly visible whenever special events and problems arise within the routine of the station, and may wish to show special support for individuals under stress during moments of crisis. Such choices and actions contribute to the reputation and endurance of the station manager.

MANAGEMENT STAFF AND ORGANIZATION

Although the manager may take principal responsibility for the creation and design of policy, policy implementation cannot be left to chance or in the hands of less than capable people. Owners realize that broadcast operation is like any other mechanism: if kept well running and maintained, it will perform well for many years, but if neglected in care and treatment, it will falter and eventually disintegrate. Staff are selected, trained, reinforced, and released accordingly.

Given the many recent changes in recruitment policies, Equal Employment Opportunity, unions, training programs, and the larger social structure, management is more diversified. Middle management and upper management still have relatively few but increasingly more women, minorities and personnel from a wider scope of previous training. The management of large staffs includes many skills and psychological, political, diplomatic, legal and motivational abilities, as well as the more traditional organizational and financial talents.

Originally, management staffing was much less complex. In some stations the owner would also be the only stockholder, serve as general manager, program buyer, and even sales manager. To some extent, this is still possible with small stations in small markets. But larger stations are usually owned by large corporations that appoint an operating general manager, often with the title of vice president. That person usually creates the various departments necessary to operate the station. It is not unusual in these larger stations for each of these department heads to also be a vice president who reports directly to the general manager. It should be noted, however, that some small stations with relatively few employees often accomplish a degree of performance equivalent to that of a larger station in a larger market, even without the latter's vice presidents and a long chain of bureaucracy. Hence, it is not the quantity of managers that determines a station's quality, but rather the quality of their leadership, relationships, programming, and understanding of their audience.

Most organizations build like a pyramid from the small capstone ownership at the top to a broad-based support staff at the ground level. Large or small, major or minor market, all television stations follow the same basic organization.

The following major elements may be arranged in various ways within an organizational chart (compare Figure 2–6 with Figure 2–2):

General Office and Operations
Sales
Programming
News
Engineering
Promotion
Research

General Office and Operations

This area accommodates many subdepartments, including accounting, billing, payroll, corporate and tax, and secretarial. The administrative staff who oversee and operate the general office are closely related to the general management. Frequently, the offices are logistically adjacent to or part of a general nerve center that includes the general management.

Operations includes the control of traffic, continuity, commercial selection, and may be a clearinghouse for the booking and scheduling of all announcements. These subdivisions work closely with the general sales manager and the director of sales to make sure that programming has proper commercial identification, and to guarantee that all materials are received. Along with sales, operations is the lifeblood of the station, because without effective operations or continuity/traffic/commercial operations, the station would cease to run.

The director of commercial operations may have an assistant who handles the day-to-day bookings and commercials. Additionally, there may be several people at a large station who handle the national level commercial operations, and several more who handle the local traffic and continuity and commercial operations. The latter team is responsible for receiving material, the booking of spots, the clearance of continuity to make sure every piece of material is acceptable to go on-air, the scheduling of tapes, the analysis of product codes so that the correct commercial runs at the correct time, and the handling of all billing discrepancies with the accounting department.

Computers are necessary for many station functions. Computers and their operators usually are part of the operations area. Some

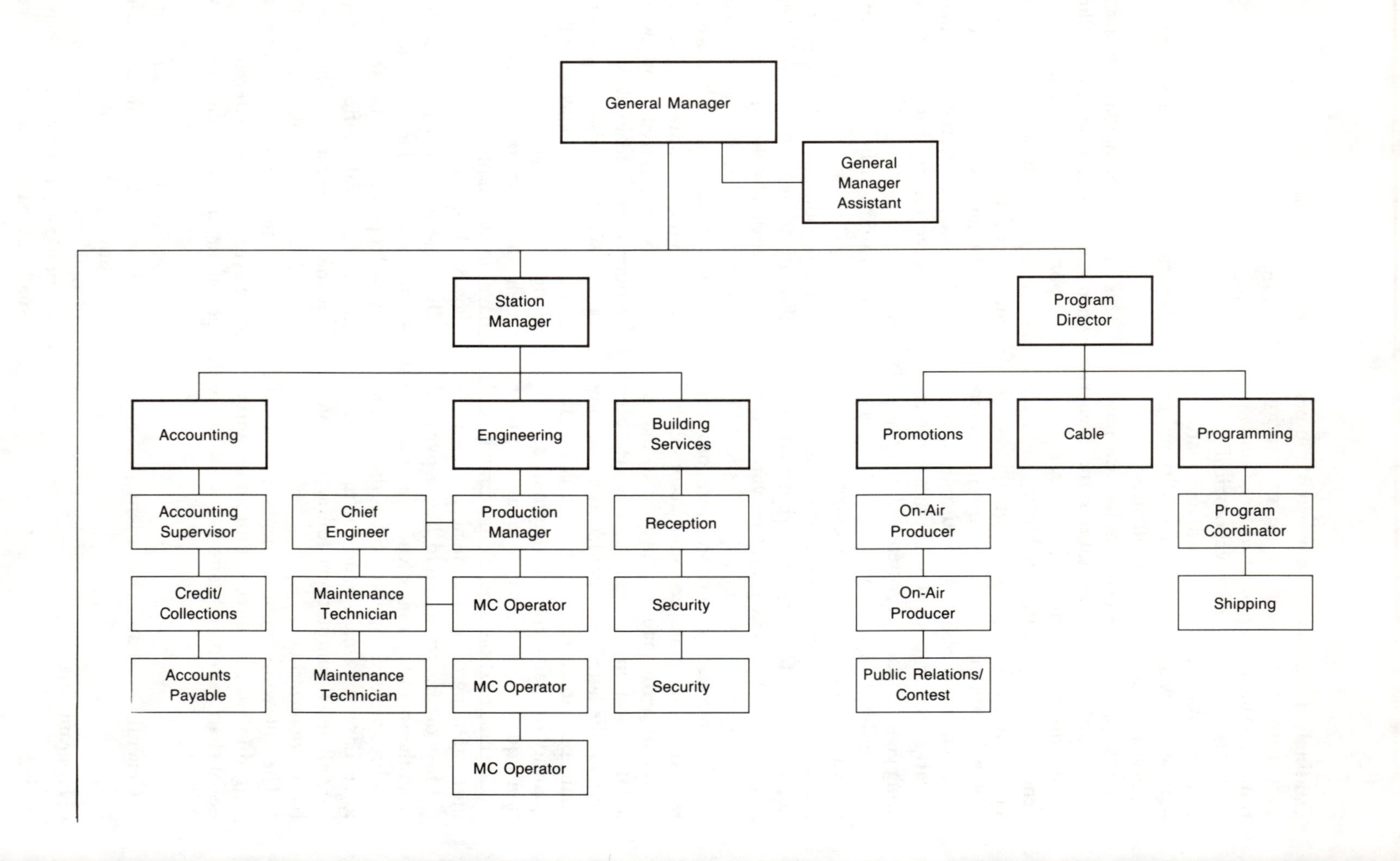
General Manager
General Manager Assistant
Station Manager
Program Director
Accounting
Engineering
Building Services
Promotions
Cable
Programming
Accounting Supervisor
Chief Engineer
Production Manager
Reception
On-Air Producer
Program Coordinator
Credit/ Collections
Maintenance Technician
MC Operator
Security
On-Air Producer
Shipping
Accounts Payable
Maintenance Technician
MC Operator
Security
Public Relations/ Contest
MC Operator

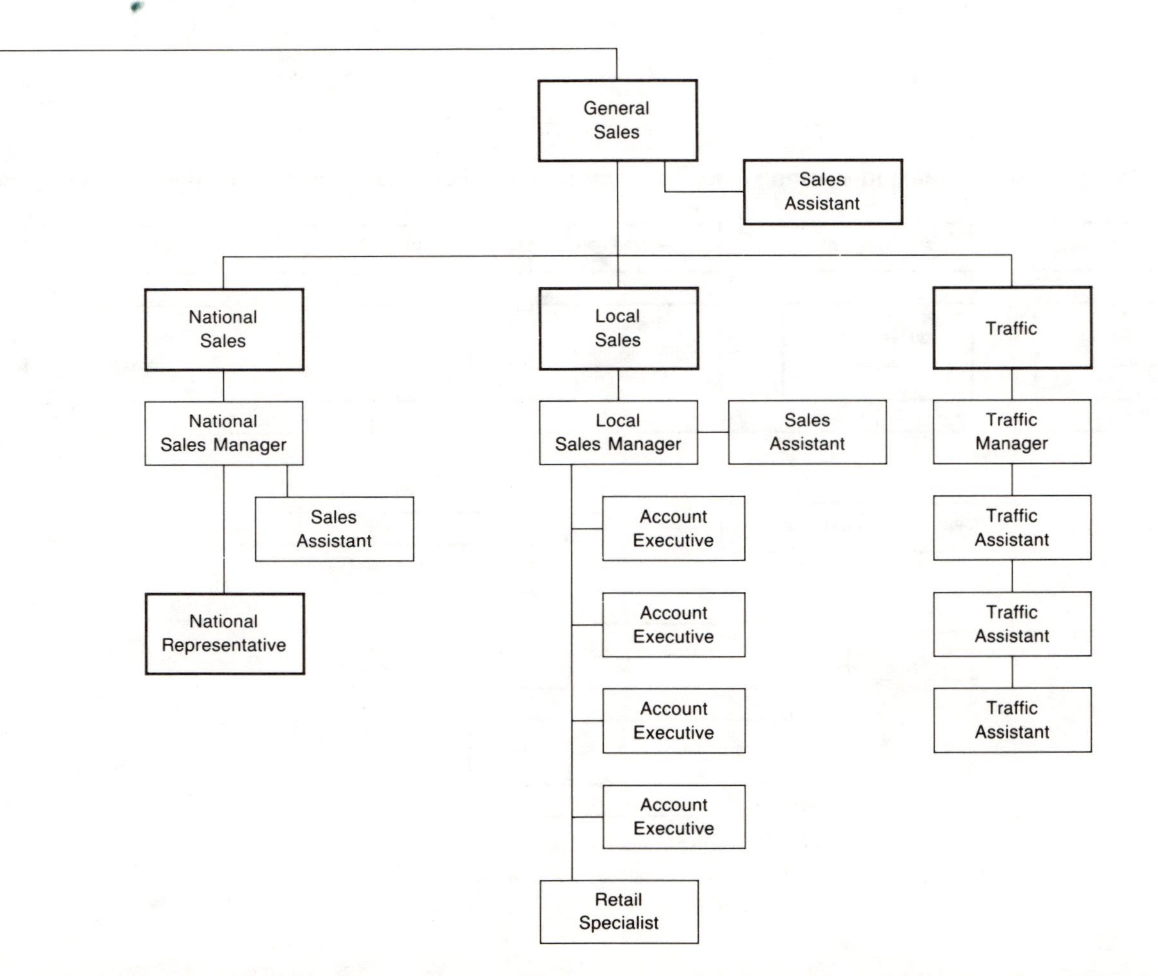

Figure 2–6A Personnel organization of a UHF (smaller) station. (Courtesy of WQTV, Channel 68, Boston.)

continued

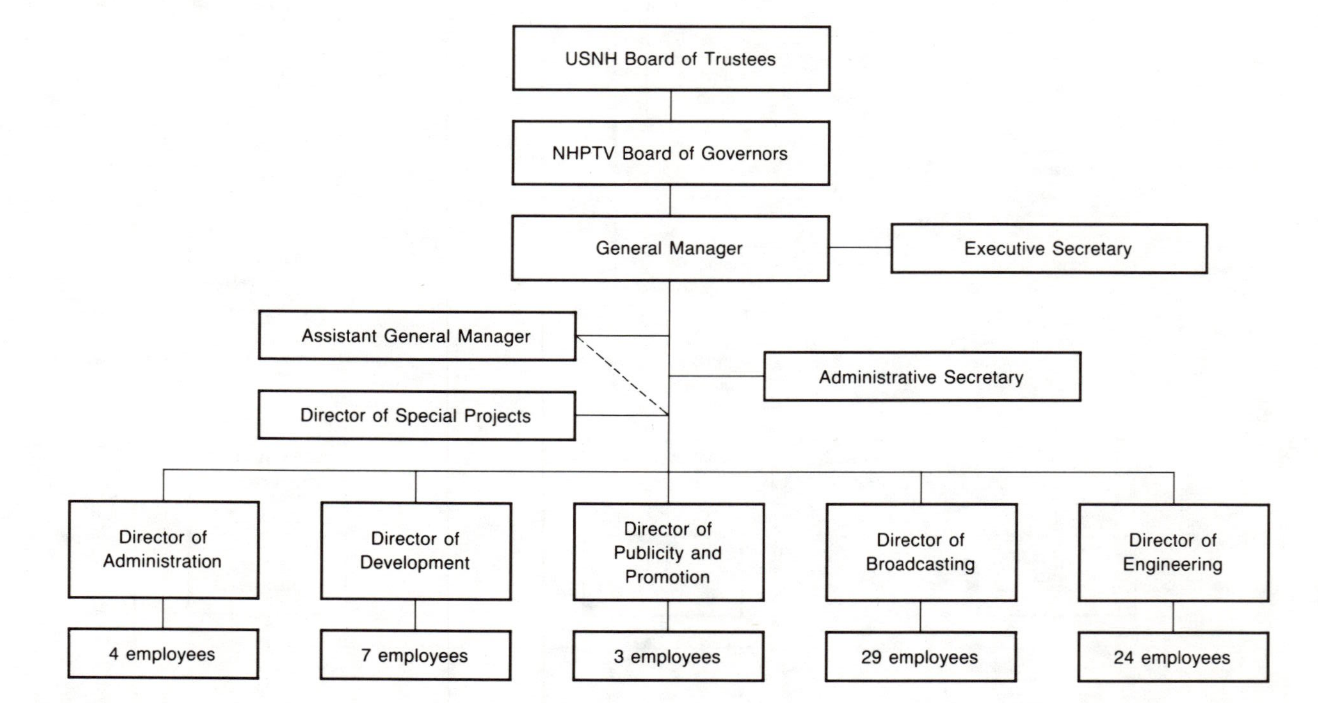

Figure 2–6B Personnel organization of a public (educational) station. (Courtesy of New Hampshire Public Television.)

of the divisions involved in computer applications are traffic (for logging and handling product codes for various advertisers), engineering (to record crew time at every event that appears on the log in the station), sales (for availability of sales records), research (for tracking audience data, trends in programming, meter data in the major markets), programming (memory of film inventory, program expenditures, and expiration dates of syndicated products), personnel (work time for all employees), and general (payroll, employee benefits, and accounting needs). As computers and computer personnel increase, it is likely that they will constitute a central and essential division or subdivision in stations of the future.

Sales

This area includes managers who are charged with sales responsibilities, including a director of sales, a general sales manager, and a local sales manager. The smaller the station, the more duties each manager within sales management carries personally. Common to all sales areas is the local sales staff (or account executives), traffic department (scheduling, commercial material, reception, logging, etc.), and the ongoing relationships with national sales clients, regional and local sales clients, advertising agencies, outside sales representatives, and other sales representatives within the station.

Many managers say that sales is the most important division of a television station. Without sales, no matter how great the programming, how thorough the news, and how exceptional the community affairs, the station will quickly die. Moreover, sales is one of the most effective training grounds for future managers. Often a young sales person, if effective, may begin in street sales, but become the local sales manager. Sometimes there is a career stopover in programming or some of the other areas to gain an understanding of the larger processes of station operations. Later he or she may move up into management of national sales, or even overall management of sales. More and more often, sales managers are being tapped for general manager slots.

Station managers work closely with national sales personnel—*national representatives*—who represent the stations in the field. Such "reps" should have loyalty to the stations they represent; on the other hand, national reps are physically distant from a given station and its general manager. National sales frequently comprise 75% of the station's revenue, while local sales produce the rest.

Sales personnel have a unique internal/external ethic. "On the street" they are highly competitive. But in the office, if one sales person is sick, injured, has a baby, or is otherwise temporarily off the job, it is an unspoken rule that other colleagues will take over the sales list, maintain it with its commissions, and save the money for the salesperson who is away. So much for the competitive, heartless stereotyped image of at least some broadcast sales persons.

Programming

The program division is one of the largest and most significant parts of a station. A program manager (also called program director) is responsible for the overall program appearance of the station.

In this time of highly competitive marketplaces, regardless of market size, the program manager and staff play critical roles in establishing the image of a station with the viewers, the community (not necessarily identical to the viewers), and the advertisers. It is no longer sufficient to offer syndicated program material and network schedules. The program manager and staff must be responsive to local citizen needs. They may also create programming that can be purchased by other stations around the country (and thus help the station's income). The key is the continual search for the right formula to increase the audience share levels of the station. While many program managers at one time came from "traffic" (the scheduling of programs), program managers today usually have creative experience as well.

The program manager supervises the placement of all programming on the schedule. With the assistance of the news director, the public affairs director, and others who contribute to the overall look of the station, the program manager guides policy as to the total use of air time.

The programming department is responsible for the on-air product. The program manager is responsible for overseeing all local program acquisition and production, contractual obligations with various program syndicators, and program relationships for the station with a network or station group.

The program department includes the staffs that produce and create local shows, talent, production and facilities schedules, writers, producers, directors, technicians not under the direct supervision of the engineering department and, rarely, but sometimes, musicians and dancers. Contracts with on-air talent, the arrangements with engineering for remote shooting (on-location shooting), and similar responsibilities fall under the program department. In a smaller station, the programming department might also be responsible for public affairs, community service, and in some instances, news. Areas such as set design and art sometimes are in the program department. The larger the station, the more programming duties are subdivided into individual departments of which department heads report directly to the general manager. Conversely, in a smaller station many of these duties are filled by a few multitalented people, each of whom fills several functions.

The program manager must be skilled in understanding research methods, sales department functions, programming decision-making, home audiences, program placement, scheduling (during prime time and other time slots), and the effects of all of this upon promotion and income. Program management must be able to use effectively the research that reports feedback on program changes and audience reaction. An effective program manager becomes expert in "counterprogramming" against the competition, judging on-air talent ability and appearance, and determining what programs are good public relations for the station. In some instances, public relations offices are under the programming department. The would-be program manager often begins as a writer, actor, assistant producer, assistant director, then moves to producer, director, film buyer, or head of traffic, and, with experience and organizational skills, subsequently to assistant program manager. An assistant program manager might move to a larger station, into a cable position, to a network, or, of course, become the program manager.

Program Selection

In larger stations the program director makes the ultimate choice, in consultation with the general manager, of the station's programming. Questions arise about whether and when to preempt network programming and carry local programs. Such decisions usually are made at the highest level, the station manager.

The program director meets film syndicators, listens and looks at their products, evaluates the programs, asks the research department for the programs' track records, consults with sales, deliberates on how the show or shows may be scheduled, and, finally, makes a recommendation to the general manager.

The program director comes to the general manager not only with a recommendation, but with an analysis of the anticipated profit margin, audience size, and evaluation of response.

Following the selection of a program another process is initiated. Contracts must be signed with the syndicator, the program must be sold to advertisers, and the promotion department edits or obtains previews for local distribution. The public relations department screens or reads copy about the program and its talent, and prepares and distributes releases to other media. Finally, an overall strategy is developed toward releasing the new program to the public, as a preview in a special debut or in some other attention-getting form.

Barter

Much programming today is barter, that is, programs supplied for airing in lieu of money, with the supplier gaining access to the station in an agreed-upon number of commercials, and the station gaining the benefit of good production and the balance of the commercial time for local station use. The program manager today must be skilled in the art of contract negotiation with the suppliers or bartering agents. If the station is a network affiliate, the manager must maintain close ties with the network for new programming arrangements, scheduling changes, and liaison between local and network programming needs.

Art

Creative design is inherent to and essential for television. In many stations, the art department operates under programming, its traditional principal function to enhance programs. Other stations, however, maintain a separate art department that works closely with the promotion department to create logos, graphics, and techniques vital for good public relations campaigns. Designers and other artists with strong commercial art backgrounds, and an ability to face extreme pressure, make graphics for news, public affairs, programming, sales, and promotion campaigns. The art department, although frequently taken for granted by viewers, advertisers, and sometimes other departments within the station, would quickly be noticed if absent.

News

The powderkeg inside a television station is often found within the news department. It is an area in which there is frequent disagreement about which programming is best. Often, viewers feel decisions about what is newsworthy are wrong. News has the capacity to inform, indict, infuriate, and insulate. News may also generate significant amounts of station revenue and at the same time cost the station sizable profit.

There are many who feel that no department within television is more maligned and misunderstood than the news division. By the same token, there are few television professionals more adulated and praised than news talent. The news director must deal with pressures from these extremes. As a manager who reports directly to the station manager, and who handles much controversial material, the news director may become one of the station manager's favorite heroes or scapegoats.

In most major stations in the late 1980s, regardless of market size, there were at least two major newscasts. In addition, Cable News Network supplied twenty-four-hour daily service, and some stations experimented with frequent but not continuous news. Evening and late evening news shows are the mainstays of most stations. Some have early morning, noon, dinnertime, and late night broadcasts. Often, the early morning programming may lead into a network program (for example, NBC's "Today Show," ABC's "Good Morning America," and CBS's "Morning News"). The responsibility of the news department does not end with the regularly scheduled newscast. News is a seven-day, seven-night per week job. Many stations have made major commitments to provide viewers with virtually all forms of news and public affairs programming. News can generate audiences, audiences deliver shares of the ratings, and audience shares deliver advertising money for the station.

As news continues to be a more vital part of a local station's programming, the production of local newscasts assumes greater importance. Even in smaller markets, local news origination is significant. Major independent stations have learned to counterprogram against strong network affiliates by running local news in alternative time slots. The usual dinner, time-to-retire, and wake-up hours are usually conceded to the network affiliates. News staffs may include far more than the production crews and studio talent for programming. As the market for news expands, payrolls include increasing numbers of production assistants, research specialists, reporters, and feature talent.

The romantic Hollywood image of tinsel and glitter in the news bureau is misleading, as is the opposite extreme of a good-hearted, heavy-drinking reporter sitting in front of an old Underwood typewriter clacking away at a world-shaking story. In the modern newsroom, feature reporters, writers, producers, talent, assignment editors, camera crews, and the news managers and directors, as well as the editorial staff, all work together in a well planned technologically orientated endeavor. Telephones seldom stop ringing. Videotape editors yell to reporters to quickly screen stories. Assignment editors dispatch crews to cover the latest event. Technical departments are upset because video levels are too low on a crowd scene shot earlier that evening. Station managers ask how the news producers will counter the competitor station's lead story that was more up-to-date. An anchorperson is caught in traffic. Sometimes, like a circus, a newsroom can be called professional chaos.

The news director usually has come up through the ranks, going back to high school yearbooks and summer jobs on hometown weeklies. He or she may have taken journalism courses, produced newscasts for his or her college radio or TV station, and spent an apprenticeship at a small station or newspaper after college. A news director may have started out reporting on the street, prepared news formats, written copy or served as an editor, or been a mid-level executive at a local paper or station prior to becoming a news director.

The skills needed are many: a sense of humor, excellent grammar, top-flight narrative ability, accuracy, fairness, and, the desire to tell all sides of a story in ninety seconds. In the larger stations, the news director is responsible for negotiating with unions, assigning staff, locating talent, reworking stories, and improving the news. The news director meets with civic, governmental, and special interest groups, works with the research department to find missing shares of the audience, and sets the budgets and priorities for the next week, month, and year.

In smaller stations, a news director may also be talent, writer, film/tape editor, reporter, public affairs programmer, or all of the above. Above all, the news director is a clear and responsible communicator who must be educated and knowledgeable in history, political science, and current affairs; constantly read the changing barometer of the environment; supervise a competent, truthful staff; and provide a living chronicle of present history. A news director must love accuracy, specificity, and responsibility, and be prepared to spend much of his or her time dealing defensively with the public, if not the station management.

Engineering and Technical Operations

Engineering includes not only the technical operations of the station, but also its technical personnel. In markets where a union bargaining unit represents the craft and technical side of the work force, the engineering manager and staff are usually the first line of interstation union relations. The engineering staff oversees all remote and local production needs. It remains abreast of technical innovations that can enhance the viewing quality and technical excellence of the station.

Cameras, time base correctors, video recorders, computers, antennae, towers, and all of the technical tools of a station must be maintained, upgraded, and smoothly operated. The coordinator of technical personnel and equipment is the *chief engineer,* who must keep abreast of the latest changes in

FCC rules and regulations, understand, interpret, and implement these changes, and know the constantly changing technology of the television business.

Math, writing, and human relations skills are prerequisities for the engineering staff. High intelligence, versatility, mechanical aptitude, and flexibility under pressure are important for people who must work with increasingly sophisticated technology. No longer can the engineering staff be content with simply running the technical aspects of the station, and making sure cameras, lights, microphones, slides, tapes, and other basic tools operate smoothly.

The technical staff is also charged with keeping both sound and picture on frequency, maintaining the signal tower, carrier wires, and antenna system, and conforming to FCC technical rules and regulations. The engineering staff today must know the coordinates for satellite feeds, the intricacies of video and audio cassette units, the application of high-density television scanning, the installation, maintenance, and operation of sophisticated new elements such as electronic still storage (replacing slide boxes), the paint box (allowing the art department to electronically create a new world of colors with multifaceted prisms), and mini-mini-cameras (allowing the operator to be a self-contained TV station by not only transmitting picture and sound, but also moving with the subjects, such as jogging with runners).

Of great importance is the chief engineer's ability to recognize the station's needs for new technology and to judge which new technological devices fit the station's size and budget.

Promotion

Most stations, even in smaller markets, have personnel who create their on-air look. Such an image may come from corporate headquarters or may be the local station's responsibility. Advertising agencies may be used for creativity-design-layout compositions and placement of print and broadcast advertising to help the station's public outreach.

In addition to overall image promotion, the promotion department works with other offices in the station on specific projects, such as long-range promotion of local programs, ongoing promotion of the news, and "tease" promotion of locally produced shows. Promotion may give feedback to the network, if the station is an affiliate, on special activities the network may do to help strengthen the public perception of the local station.

Promotion sometimes produces on-air spots and creates material with and for community groups. Often, the promotion department creates teases or lead-ins, short previews of upcoming programs that seduce viewers to stay with the channel or to tune in again later. Outreach promotion may also include billboards, signs on park benches, promotional contests, and even sky-writing and movie trailers.

Research

Research plays such an important role in television that at many stations it is a separate department or a large, semi-independent office within the sales division or the programming area. Sales and programming research play a significant role in determining format, advertising time costs, and program placement. Research is a specialized job that requires understanding of the marketplace, the competition, the station itself, and experience in other areas of the broadcasting business. Research personnel work closely with the rating services, Arbitron and Nielsen. They purchase or create demographic studies and compile information for individual station departments, including news and public affairs, and promotion, as well as sales and programming.

The larger the station, the larger the research staff. Networks maintain large research departments, some for future planning, others for present analysis, and still others for determining past trends to prognosticate the future.

LEGAL AND REGULATORY CONSIDERATIONS

Operating a television station requires great understanding of and compliance with many laws and rules. Whether the FCC is increasing regulation or is in the process of deregulation, management must be in step with the latest changes. To violate FCC guidelines can be costly to a station's reputation, financing, and even operation. A license may be revoked or suspended, and the station can be fined or reprimanded. In most cases, the cost to a station is in management's time and in lawyers' fees.

However, it is not only the FCC that has governmental influence on the direction of the television station. The Federal Trade Commission (FTC) serves as a sometimes watchdog on advertising not in the public interest. Federal laws, such as the Equal Time rule in political campaigns, affect TV operations.

Many people are not aware of the complexity of other types of regulation that must be followed by television management. Consumer laws, for example, protect the public against products that have been banned. Federal banking laws are important, affecting a station's tax structures. Most physical changes for a station—increasing the power, raising the height of the antenna, modifying the building, changing the ground system—relate to city and/or state zoning or environmental impact laws, as well as federal regulation. Even with deregulation, a TV station is still required, at this writing, to renew its license every five years, and although public participation in the process was dramatically decreased in the 1980s, there is still some opportunity for the pubic or a competing potential licensee to challenge the station, thus maintaining some required measure of public interest and public service in broadcasting.

COMMUNITY CONSIDERATIONS

When the news reaches a community that a broadcast station is seeking to move into its area, many reactions are possible. At one extreme is the welcome mat approach. Many communities know that a successful station will contribute significantly to a tax base, bring revenue to the city budget, and promote business and commerce. Communities know that television can enhance a town's public relations, offer media access and outreach to civic representatives, provide forums for political candidates, and even supply a decorated television tower during the Christmas holidays.

At the other extreme, a community may perceive a television station as a threat to the media status quo, an eyesore on their landscape, unnecessary modernization for a traditional community, or a disruption in some other way. A community that does not welcome a television station may find many ways to prevent its location and growth, including zoning laws and other legal restrictions.

The attitude of the station toward the community is equally significant. Most stations advocate an active good neighbor policy. Station management realizes that the station and community must cooperate if the station is to establish a respectable audience and, consequently, a respectable income. A reciprocity psychology underlies this approach: as the station treats the community, so the community will treat the station. Management should be open-minded to community affairs, and carry as many local events as it can. It is good business for management to support and spotlight community activities.

Many stations are more network oriented than others and decide that community programming is not profitable. Local programs usually are considered artistically inferior to high-budget national products. At one extreme are stations that seek to serve the community wholeheartedly; at the other extreme are stations that wish only to give the appearance of serving the community's needs, but that, in fact, provide the minimum community service possible.

Some larger stations, such as WCVB-TV in Boston, have a separate community relations department, as large as the entire staff of several local radio stations. WCVB entered a marketplace in 1972 where there was little more than ten hours of community programming on the air. When WCVB signed on the air it brought thirty hours per week of its own local programming. By 1985, one third of the station's scheduled programming was locally produced.

Successful station managers make certain that they, individually, and their staffs become part of the community and participate, as citizens, in clubs, associations, and other civic endeavors, as well as seeing to it that the station, outside of overt advertising gain, sponsors and otherwise assists the citizens in events of importance to the community.

MANAGEMENT ETHICS AND CHARACTER

Ethics

The reputation of a station manager and staff reflect upon the reputation of the station. The ethics or moral characters of a station's employees may reflect negatively or positively upon community relations and professional image. Specific ethical issues such as libel, invasion of privacy, and protection of sources must be understood by management, especially by the news director and station manager. However, the larger context of social ethics is also important for all management.

Social ethics relates to the citizenship of a station's members. The station itself is a small city, often with its own cafeteria, emergency employee accommodations (e.g., during a snowstorm), printshop, and even a flagpole. However, it is a city within a larger city or county government. To the degree that members of the smaller "station city" are respectful and responsible to other members of the larger surrounding city or county, will fluid cooperation among the two entities exist. Each entity has a type of self-government, but the station's self-government is responsible to the larger local, regional, and federal governments. Thus, the ethical practices of station management influence larger civic groups, viewers, advertisers, and other organizations.

Those working within television, like those working in politics, are under close scrutiny. Drug problems, domestic spats, alcoholism, and other personal difficulties may be more readily publicized than those of less visible citizens. The fishbowl-type existence of television management and talent calls for great personal care, since any small behavioral aberration may be grossly exaggerated through public prominence.

Character

Some of the conscious qualities that must be developed by a skilled manager have been articulated: financial wisdom, group psychology, disciplined organization, programming expertise, cost-effectiveness, and operational consistency. But there are many more subconscious traits that managers communicate by means of attitude. The perceived character of any individual is an influential force that may either inspire or alienate other members of the station team.

Specific qualities of character are essential in television management. For example, perseverance is a necessity: without the ability to follow-up, even in the face of possible failure, a person will not maintain his or her current position, let alone rise to higher management levels. Patience is a complementary quality to perseverance. While the caricature of television managers who lose their tempers over any trivial detail has some basis in reality, managers who walk out on problems, who "blow their cool" and melt under pressure, have much lower survival rates than those who hold steady, bounce back, wait out the storm, and solve the problems.

Television stations can be likened to pressure cookers. Consequently, those aiming toward management positions must learn first to handle pressure and eventually to harness pressure. The handling of pressure is important so that deadlines and periods of intensity do not stimulate erratic responses among personnel. The true leader learns to welcome pressure as a necessary and creative component in the refinement of a product. Knowing how much pressure to apply, at what time, and to which department or individual in order to increase productivity or counterbalance a departmental nose dive is an essential yet delicate skill. When correctly harnessed, pressure may be seen as benevolent and, indeed, the motivating factor in an organization.

Sensitivity is essential for successful management. A personal antenna that will pick up early warning signals, detect individual burnout tendencies, spot areas of exceptional stress, and pinpoint necessary encouragement patterns with specific people must be used. The subconscious sensing of the motivations, thresholds, and pivot points for other people guarantees a greater depth of success. Management must be sensitive to the perceptions not only of employees and colleagues, but of the community, advertisers, and official agencies. Observing and correctly translating the emotions, perspectives, and goals of others is vital for the effective manager.

Leadership is ineffective without teamwork. A responsible manager must learn the secrets of motivating and developing other managers and, indeed, all personnel. Dormant talents should be encouraged to develop. There is a thin line between being supportive of an individual's potential and pushing the individual beyond his or her personal threshold. Moreover, there is an equally thin line between disciplining colleagues with problems and breaking their spirits entirely through authoritarian behavior. A wise manager must learn to understand the relative metabolic and emotional dispositions of the staff and, like a talented orchestra conductor, help facilitate a harmonious blending of the parts of the organization.

Beneath these psychological skills is a less analytical process. A successful manager must genuinely love people. If motivational techniques and feigned empathy seem artificial, no real leadership is expressed. Underneath all techniques and tendencies must be a concerned "people person." A manager will always be surrounded by people; television leadership jobs are not for those with hermetic tendencies. The authentic concern for people will be detected not only by employees but by community leaders, advertisers, representatives of other industry institutions, visiting talent, and, indeed, the entire galaxy of individuals upon whom the station is dependent. Thus, the qualities of authenticity, honesty, and trustworthiness are essential for the manager who wishes be successful.

BALANCING RESPONSIBILITIES

Among all the qualities and responsibilities of effective management listed above, some rate higher than others. Certainly, strong personnel relations, programming skills, and protection of the bottom line are essential to the success of any station. Within personnel relations are many specific practices such as the provision of employee benefits and salaries, humane work conditions, wise policy, early implementation of new technology to facilitate work responsibilities and achieve cost efficiencies, and general courtesy and understanding. Protection of the bottom line requires other specific skills such as efficient budget management, careful prognostication, understanding of departments and specific operations, legal knowledge, and particularly good business common sense.

Size of market and of staff never guarantee success. The people who make up the staff and their contributions make crucial differences between monetary success and failure. Financial sensibility, however, is critical. For example, wise management will institute effective cost-watching procedures to ensure the profit and stability of staffing. Moreover, wage and salary guidelines must be followed and union contracts taken into account. A large vision is important to anticipate all financial variables. Good legal counsel must be intelligently incorporated to deal appropriately with the myriad of local, state, federal, and even international rules and regulations.

Management skills must be flexible, and the manager must always be willing to learn new ones. Teletext, satellite transmission, cable operations, DBS, HDTV, laser beam technology, home computers, office computers—all of these and more must be understood by management in the 1980s.

Beyond technology and economics, however, is the one key element that makes a television station work: the people. No amount of exotic promotion, razzle-dazzle equipment, or sophisticated education can substitute for the experience, attitudes, and effectiveness of solid television management. While equipment, technique, and promotion certainly assist a station that is already well endowed with effective people, nothing can save a station in which individuals are incompetent, unethical, and resistant to change. People central to television operations, whether at management or entry levels, must possess or develop specific attitudes and personal qualities suited to high stress, multiple deadlines, and constant uncertainty.

Technique and talent, although important attributes of strong management, are nevertheless secondary. More important are the wise handling and harnessing of pressure, the ability to motivate and understand a wide variety of people, honesty within a profession closely watched, and the consistent stamina required to surpass, if not transcend, cutthroat competition.

The real bottom line means more than protecting the financial bottom line. Developing top line leadership that will attract top-of-the-line staff and talent is vital. Business tenacity and protection of profit are crucial. But without a team of genuine, flexible, and effective people, such tenacity and protection may prove worthless. Responsible leadership and coordinated teamwork are the backbone of success, not simply for the station and for management, but for communication and for society.

BIBLIOGRAPHY

Coleman, Howard W. *Case Studies in Broadcast Management.* 2d ed. New York: Hastings House, 1978.

Compaine, Benjamin M., et al. *Who Owns the Media?: Concentration of Ownership in the Mass Communications Industry.* 2nd ed. White Plains, NY: Knowledge Industry Publications, 1982.

Cooper, Thomas W. *Television and Ethics: A Bibliography.* Boston: G.K. Hall, 1988.

Ettema, James and Charles Whitney, eds. "Individuals in Mass Media Organizations: Creativity and Constraint." In *Sage Annual Review of Communication,* no. 10. Beverly Hills, CA: Sage Publications, 1982.

Levine, John M. and Daniel Wackman. *Managing Media Organizations.* New York: Longman, 1986.

Marcus, Norman. *Broadcast and Cable Management.* Englewood Cliffs, NJ: Prentice-Hall, 1986.

McCavitt, William E. and Peter K. Pringle. *Electronic Media Management.* Boston: Focal Press, 1986.

Quaal, Ward L. and James Brown. *Broadcast Management.* 2d ed. New York: Hastings House, 1976.

Reed, Maxine and Robert Reed. *Career Opportunities in Television and Video.* 2d ed. New York: Facts on File, 1986.

Shearer, Benjamin F. and Marilyn Huxford. *Communications and Society: A Bibliography on Communications Technologies and Their Social Impact*. Westport, CT: Greenwood Press, 1983.

Victor, Richard and Davis Dyer, eds. *Telecommunications in Transition: Managing Business and Regulatory Change*. Boston: Harvard Business School, 1986.

3

Sales and Advertising

Robert L. Hilliard and Kathleen P. Mahoney

THE FRAMEWORK OF COMMERCIALLY SUPPORTED TELEVISION

Before a television station can begin to serve its community with quality programming, the management first must secure the necessary financial resources. This responsibility falls to the station's sales department. For this reason the sales department is "first among equals" within the organizational structure of the departments of a station. Sales departments of network-affiliated stations bring in about 95% of the revenues through commercial sales time. At independent stations, sales departments bring in 100% of the revenues through both commercial and program time sales.

It is difficult to believe today that in the early days of broadcasting advertising on the air was frowned upon. In the early 1920s, at the time the first commercial was being aired on New York radio station WEAF on August 28, 1922, Secretary of Commerce Herbert Hoover, whose Department ostensibly had jurisdiction over radio, argued against advertising on the airwaves. It was proposed that broadcast stations derive their revenue from sources such as a receiver sales tax, a licensing fee on receivers (as is the practice in many countries in the world), or direct audience contributions. Interestingly enough, it was the same Herbert Hoover who appeared in the first test of television, a hookup between Washington D.C., and New York City on April 7, 1927, an event reported on the front page of *The New York Times* the next day with a subheadline, "Commercial Use In Doubt."

Had a no-advertising approach prevailed, the structure of broadcasting and the content of programming today would be quite different. Broadcasters quickly discovered, in the early 1920s, that advertising was the most profitable source of income. The commercial broadcasting industry today is based on advertising.

Advertising plays a significant role in determining what is aired on television. In analyzing the interrelationships among various participants in mass media, Joseph Turow combines resource dependence and media power roles to illustrate the impact of advertising on programming. In the mass media advertisers are external sources of power.[1] Within media organizations, especially broadcasting, the sales department is the liaison with the station's advertisers. Concomitantly, the sales department translates that power into its internal role in the station. How does that translate into programming? Turow argues that television executives create schedules, especially in prime time, that will attract the demographic group most sought after by advertisers: women eighteen to forty-nine years old.[2] Television broadcasters rarely air so-called narrowcast programming, as do radio stations, because of the need in television to attract large audiences. The nature of radio, since its national appeal was displaced by television, has been specialized local programming to selected narrow slices of the audience pie. The sales department keeps management informed of the advertising environment: market conditions, advertiser demand, and audience trends.

That is not to say that sales directors dictate programming. Schedule decisions are made by the general manager in consultation with many department heads. Numerous concerns are taken into account when making such decisions, including the "look" of the station and its service to the community. Nevertheless, the advertiser will place commercials only on the kinds of programs that will attract large numbers of the kinds of audiences the advertiser wishes to sell to. The role of the sales department in programming is to inform management of advertiser demands and to represent the fiscal perspective in management decision-making.

The principal goal of the sales department is not managerial. It is to secure station revenue. Without this revenue the other departments would have no operating resources. In the resource dependence model, the sales department obviously exercises direct power internally over dependent departments.

ORGANIZATION OF A SALES DEPARTMENT

No standard format exists for organizing a television station's sales department. Differences among departments occur because of market size, network affiliation or nonaffiliation, dominance of a station within its market, and differences in management style. Figure 3–1 illustrates the basic job positions found in most sales departments.

All departments are headed by a director of sales or, as the position is also called, a sales manager. In some stations the job is designated as vice president for sales. This position reports to the general manager of the station. Two important positions for managing time sales report to the director of sales: the national sales manager and the local sales manager. In small and medium markets the director of sales and the national sales manager frequently are the same person. The local sales manager manages a staff of account executives or, as they are also called, salespersons. The number of salespersons at a station varies widely and is dependent on the place of the station in its market and whether or not the station has a

Figure 3–1 Basic organizational chart of a sales department.

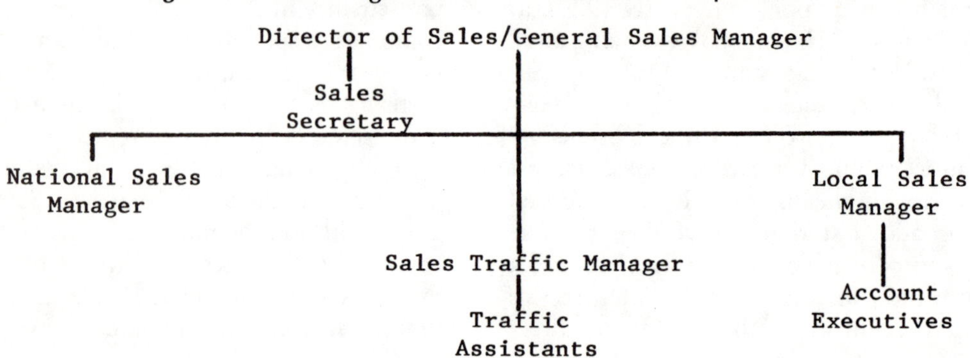

network affiliation. For example, in Binghamton, New York, market 133 in the country, the dominant station is the sole VHF station in that market. This station has seven local salespersons while a UHF station in the same market has just three. The number of local salespersons does not necessarily vary by market size, however. Binghamton, Syracuse (market 66), Buffalo (market 34), and Philadelphia (market 5) all have about the same number of local salespersons at network-affiliated stations. These stations have much inventory to sell and therefore need large sales staffs.

The sales traffic manager, noted on the chart in Figure 3–1, is not found in all sales departments. Some stations place sales traffic and program traffic in separate departments. There are a number of other positions that vary in different sales departments, including regional sales manager, new business development manager, and audience research director. At stations where there is a large volume of program and commercials production, for the station's clients and for others in the area, the sales department might have a commercial production unit.

Independent stations often have two additional positions, program sales manager and sports sales manager. The program sales manager brings in paid program-length time sales. For example, some religious and some self-help programs fall in this category. No time is sold within the program and the producer pays the station for the entire time the program runs. Many independent stations hold sports broadcasting franchises from local and regional athletic teams, and the job of the sports sales manager is to negotiate time sales for the sports program.

Interdepartmental Relationships

While the exact positions within the station sales departments may vary greatly, the basic function of departments within stations is remarkably similar. No department in a station operates in a vacuum, least of all sales. Even as other departments are dependent on the success of the sales department for their existence, the sales department depends on other departments to help it carry out its functions. Sales most frequently interacts with two areas, production and accounting. Production of commercials generally does not fall within the control of sales. In order to get client commercials produced, account executives (salespersons) must work with the production department. Interactions involve all members of the sales department at some point, requiring a solid working relationship.

Although account executives work closely with accounting to determine credit histories on their clients, the more significant relationship is between the director of sales and the head of accounting. Much of the sales director's job relates to financial forecasting; that is, what the expected sales revenues are. In order to determine the bottom line, the director needs continuing current financial data from accounting. Free, uninterrupted information flow between the two departments must be established.

Sales also interacts frequently with programming and promotion. The sales director and program director (sometimes called program manager) both provide input on prospective program purchases to the station manager. Conflicts occasionally (even frequently) arise. The sales department advocates program purchases that it can most effectively sell to advertisers. The programming department is concerned with the image of the station in the community and in promoting good relations between the station and its program suppliers. In the event of conflict, the station manager decides on the basis of overall station interests and the policies of the corporate owners.

The promotion department interacts in a very different way with the sales department by providing the salesperson with well designed materials for sales presentations. Special events are sometimes organized by the promotion department. For example, clients may be invited to the station for a luncheon preview of new programming. The promotion department frequently selects one or more programs for promotion in other media and through special events in the community at large, as well as promoting the station image as a whole in such ways. A key requirement of a promotion department is to establish the station as an integral and contributing part of the community, thus promoting more viewership and more goodwill among potential advertisers and the latter's potential customers, in this way directly facilitating the job of the sales department.

Most interaction, then, between the sales department and other offices in the station, is with these four departments: production, accounting, programming, and promotion. Sales managers at most stations are careful to stay uninvolved with the news department, fearing that they might give the appearance, even if not actually doing so, of influencing the type or extent of news programming in order to please the station's advertisers. The sales department must know the news inventory as precisely as it does other programming because it must sell commercials within its news programs as well as for other programs. But it must remain clearly apart from decisions affecting news content.

Cable Systems

Cable system sales departments operate virtually the same way as do local sales personnel at broadcast stations. At this stage of their growth local cable systems do not seek or garner much national advertising through national sales representatives, as do local broadcast stations. Instead, however, because they offer specific local audiences to local advertisers, and in smaller numbers, they have rates much lower than broadcast stations in the same community and are able to draw new advertisers who have only minimal amounts of money to spend on the media. In addition, they can offer time periods frequently at the sponsor's wish or whim. In the mid-1980s, about 50% of the nation's homes were wired for cable. Some estimates show that growing to about 70% by the early 1990s.

Some local cable systems have regular and growing sales staffs and in some high-penetration markets cable systems are posing direct competition to stations. Cable sales in many communities are beginning to fill local origination programs and in some instances even public access programs; that is, the programs originated in the community or obtained from other sources and sent out from the local system to the local community. Other programs, those on the broadcast stations or on special or premium cable services the local system carries, already have commercial time inserts sold by the source. Many national cable program distributors operate in sales similar to broadcast networks and obtain national advertising for their programs.

NATIONAL AND LOCAL SALES

Stations can have very skilled sales staffs but still be down in sales. Factors beyond the control of station management can affect both national and local advertising sources. Weaknesses in each of these areas are produced by different variables. Top station management must understand and anticipate market trends in these areas in order to set station budgets. The sales manager must take these trends into account in setting and negotiating rates and in developing new and existing client orders.

While national and regional spot sales consistently provided more income for stations than did local sales, in the 1980s the trend began to move the other way, and some prognostications indicate that by the 1990s revenues from local sales will exceed those from spot sales. Figure 3–2 illustrates a thirty-year breakdown of network, spot, and local sales.[3]

Year	Network compensation		National regional spot (In Millions)		Local	
	Amount	% change	Amount	% change	Amount	% change
1963	$203	0.5%	$616	11.2%	$241	5.7%
1964	214	5.4	711	15.4	276	14.5
1965	230	7.5	786	10.5	303	9.8
1966	244	6.1	872	10.9	346	14.2
1967	246	0.8	872	0.0	365	5.5
1968	248	0.8	998	14.4	453	24.1
1969	254	2.4	1,108	11.0	519	14.6
1970	240	−5.5	1,092	−1.4	563	8.5
1971	230	−4.2	1,013	−7.2	637	13.1
1972	224	−2.6	1,167	15.2	778	22.1
1973	233	4.0	1,221	4.6	896	15.2
1974	248	6.4	1,329	8.8	979	9.3
1975	258	4.0	1,441	8.4	1,080	10.3
1976	270	4.7	1,920	33.2	1,390	28.7
1977	288	6.7	1,960	2.1	1,586	14.1
1978	315	9.4	2,326	18.7	1,987	25.3
1979	344	9.2	2,564	10.2	2,245	13.0
1980	369	7.3	2,920	13.9	2,484	10.6
1981	393	6.5	3,302	13.1	2,767	11.4
1982	406	3.3	3,846	16.5	3,088	11.6
1983	416	2.5	4,211	9.5	3,611	16.9
1984	424	1.9	4,715	12.0	4,216	16.8
1985	446	5.2	5,077	7.7	4,665	10.6
1986	454	1.8	5,574	9.8	5,275	13.1
1987	460	1.3	5,795	4.0	5,610	6.4
1988	469	2.0	6,350	9.6	6,325	12.7
1989	476	1.5	6,825	7.5	6,900	9.1
1990	483	1.4	7,275	6.6	7,475	8.3
1991	489	1.3	7,725	6.2	8,025	7.4
1992	497	1.7	8,525	10.4	9,025	12.5
1993	504	1.4	9,150	7.3	9,800	8.6
5-year growth rates:						
	(1975–1980)	7.4%		15.2%		18.1%
	(1980–1985)	3.9		11.7		13.4
	(1985–1990)	1.6		7.5		9.9
10-year growth rates:						
	(1975–1985)	5.6%		13.4%		15.8%
	(1980–1990)	2.7		9.6		11.6
5-year growth rates:						
	(1988–1993)	1.5%		7.6%		9.2%

Sources: FCC television financial data (1963–1980); TELEVISION/RADIO AGE "TV Business Barometer" (1981–1986); Dick Gideon Enterprises (1987–1993)

Figure 3–2 Sales of United States television stations by advertising category, 1963–1993. (Courtesy of *Television/Radio Age*.)

National Sales

National (and regional) sales are composed of the *spot* market and barter syndication. The term *spot* as used here does not refer to an individual commercial, as in commercial spot, but to the spotting of commercials in selected local stations throughout the country by the advertiser. Many advertisers use the spot approach alone and do not make any network buys. Spot advertising is taken in directly by the station and is of primary importance in understanding the operation of national sales. Placement of spot advertising, in the opinion of many experts, is the result of so-called spillover once the networks are sold out. The advertisers who have no more network availability then place their ads on local stations, although some experts, such as William Schrank, senior vice president of Katz Communications, argue that this is not necessarily the case.[4] National spot sales are a significant part of a station's income.

The sales staff is dependent on the rating charts in order to sell time, both nationally and locally. The larger the audience a given show garners, the larger the cost per 60-second, 30-second or 15-second (split-30) commercial in and around that show. Costs vary for each program and even for placement within the program (i.e., before the opening, in the first or second break, before or after the final credits).

The Neilsen ratings, for example, let prospective advertisers know what programs and what stations are getting what percentage of TV households and what percentage of sets tuned in at given times. The advertisers, whether national, regional, or local, then know where to best place their ads. The Neilsen ratings measure 200 markets four times a year, with the largest markets sometimes measured several additional times during the year. Arbitron, using meters as Neilsen does, and diaries in selected households reinforced by phone calls, provides local demographic information to its clients. Partially in response to long-time criticism of the validity of systems that measure principally whether a set is on and not whether anyone is watching, Arbitron in 1987 began using a "people meter," with a heat sensor that determines whether there are people actually within viewing range of the television set.

National spot markets are subject to the individual decision-making processes of advertisers. William Schrank says that the impact of such diversity is unknown. What are some of the decisions individual advertisers take into account? Some use spot as a last-minute buy. Such decision-making is idiosyncratic and difficult to predict. Many advertisers have begun to rebel against the high cost of television, especially network, and have sought either alternative media or price reductions. In studies done by the Ted Bates Agency, television advertising exceeded both inflation and other media in yearly cost growth over a ten-year period, 1976–1986. Spot television was slightly ahead of inflation for those years. Spot radio, outdoor advertising, and all forms of print were much more conservative in cost increases to advertisers for this period.[5] Advertisers generally have not reacted to the trend towards higher television costs. However, individual advertisers have reacted, just as individual advertisers use spot in different ways. Overall, the impact on spot may be gradually affected by these factors.

Just as local markets are affected by retail sales, so are national markets. According to Schrank, this is the only factor known to directly affect advertising sales. For example, in the 1980s Procter and Gamble was by far the leading advertiser in both spot and network markets. Nationally its retail sales remained consistently strong, but individual markets saw declines because of changes in the local economy. Specifically, in the early 1980s some northeast markets were in decline, while many southwest markets were experiencing a financial boom; by the mid-1980s the southwest oil-industry-based markets were experiencing high unemployment and large population migrations while northeast markets regained strength. Spot advertisers seek out markets where consumer disposable income is high or where favorable demographics prevail.

Nationally, spot can be strong, but on a market-by-market basis it may show weakness. This can be the result of either temporary market decline, as discussed above, or the result of inherent market characteristics. Spot is a greater portion of station revenues than local advertising in larger markets where the population is concentrated, the television market is competitive, and the economy is varied. In very small markets (150+) spot dominates station revenues since fewer advertisers are available in these markets. Within the middle markets spot generally represents less than half the station's revenues.[6] Within these markets some stations do attract significantly more spot either because they have diverse demographics or highly sought after specific demographics.

Network National Sponsorship

Different than spot advertising, the networks negotiate national sponsorship, whether program, coop, alternate, or single commercial (the other definition of *spot*). The local station affiliate negotiates with the network for time within a given program for local commercials sold by the station and for station identification. All affiliates of a given network are given the same local origination time. In a popular program they can charge high rates to local advertisers. Affiliates are also compensated by the network for carrying network programs by being given a piece of the national advertising pie. In the mid-1980s this averaged about 10% of the compensation received by the network. This varies in terms of time of day: prime time pays more, daytime shows less. Sports events usually pay nothing to affiliates because of the high cost to the network of acquiring broadcast rights. However, local stations can do well with time sales adjacent to the sports event and, as negotiated with the network, local origination time within the event.

Barter

There are two principal ways in which a local station can get some remuneration for time that ordinarily wouldn't be sold: *barter* and *tradeout.* Barter is usually done on the national level, where commercial time is exchanged for a product or service, usually something of significant value, such as automobiles that may be used by the station or European vacations that can be given away as station promotion prizes. Tradeout occurs on the local level and will be discussed under local sales.

Barter syndication has been a point of conflict between stations and syndicators in recent years. Even the top shows may have barter components. For example, "The Cosby Show" syndication was sold for record cash prices plus barter spots within the program. Barter began to grow in the 1980s and has become a major force in the marketplace, as illustrated in Figure 3–3.

Many local stations have been concerned that barter syndication has an adverse impact on overall national spot sales and on station inventory. Does barter syndication hurt spot sales? The data are not able to resolve the conflict. Advertisers continue to deny that barter is a substitute for spot. Studies done by the Station Representatives Association and by Lorimar-Telepictures argue that it replaces network buys.[7] According to Schrank, even if barter does take dollars from network, this can still hurt spot sales. In a weak market, networks may drop prices, making it more affordable to advertise on network. Surveys of station managers for the Station Representatives Association and discussions with sales managers indicate that most believe that barter syndication hurts the local station's spot opportunities.

Also, by giving up spot sales to barter, the station loses inventory that could be sold either on the spot or the local market. The impact on inventory is virtually impossible

	Spot		Barter Syndication	
Year	Millions	% Change	Millions	% Change
1980	3,269	13.8	50	—
1981	3,730	14.1	75	50
1982	4,360	16.9	150	100
1983	4,796	10.0	300	100
1984	5,453	13.7	430	43.3
1985	5,950	9.1	540	25.6
1986	6,570	9.4	610	13

Source: Television Bureau of Advertising.

Figure 3–3 Television ad volume, 1980–1986. (Courtesy of Television Bureau of Advertising, New York.)

to clearly determine because stations rarely sell out. The value of barter spots to the stations is unknown since one cannot determine how much of this inventory the station would have sold. While syndicator George Back argues that barter helps stations to sell inventory,[8] sales managers argue, on the other hand, that any inventory that is taken from the sales rolls hurts the station.

The arguments over barter syndication will not quickly subside, although in the late 1980s barter was becoming less attractive to advertisers. The price of syndication was nearing network pricing; syndication was saturating the market, making clearances more difficult; and the growth of independent stations was leveling off, thus leveling the outlets for syndication. As barter becomes a more stable factor in the market mix, its impact on spot may be clearer.

Station Representatives

For national and regional spot business the local station seeks advertisers who happen to be mostly, if not all, outside of the station's community. Therefore, the station needs an advertising representative or firm to make contact with the advertisers and with their advertising agencies that prepare and market the commercials to the appropriate slots. Station representatives negotiate with advertisers on behalf of the local stations. The *station reps* keep the prospective advertisers, nationally and regionally, informed of the rates and time availabilities of the stations they represent. Station rep firms are growing, more and more replacing individual sales persons from stations, with their national and regional contacts and expertise. By being able to consolidate information, and using computerized data bases, station rep firms provide sophisticated research support that most small stations can not do on their own.

The rep firm is employed by the station under an exclusive market contract. Therefore, although a firm may represent many stations, it usually has only one in any given market. The station rep must sell national and regional spots on the basis of demographics, just as does the local salesperson. It is important for the firm to continually research the market it represents, and its station's place in that market. The sales manager must constantly keep the rep informed of audience research, format changes (which, of course, are designed to increase sales), and any other factors that might facilitate the rep's ability to sell the station to advertisers. Frequently, when the station is contemplating changes in programming or promotion, it will consult with the rep before finalizing any plans. Rep firms frequently provide program data, analysis, and advice. This may include national and local analysis of program performance as well as analysis of program purchasing potentials. By ensuring the best programming on client stations, the reps are ensuring better sales and revenues for their own firms.

By saving the advertiser the chore of contacting individual stations throughout the country, the rep also provides a service to the advertiser. However, it is the station that pays the rep's fees. Commissions vary by market size: less than 10% for a large market station, 10–12% for medium markets, and as much as 15% for small market stations.

Because some station representative firms handle many stations, the individual station's sales manager must keep in touch with the rep on an ongoing basis to find out progress on sales, to inform the rep of any new demographics or other developments that may make the station a more desirable advertising medium, and to simply motivate the rep to work harder for the station.

Local Sales

Just as many market factors can influence the spot market, local sales are also influenced by some factors out of the control of stations. Retail sales are the key to local sales revenues. Market decline spills over into station profits. However, advertising dollars do not stop flowing into stations. Declines often take place after a boom period. Television sales may stabilize at a level of 10–12% increases yearly compared to higher yearly sales increases, often in the 20% range. Most large markets can bear a general decline because local economies are diverse enough to recover.

In addition, individual markets vary widely in the types of advertisers that comprise major clients. CBS Vice President for Research David Poltrack compared the top twenty advertiser categories for New York and St. Louis. Of the twenty, six do not appear on both lists. Looking at individual advertisers, only two appeared on both lists.[9] A decline within a retail category may affect some markets but not others. Local market sales exhibit a great deal more idiosyncratic characteristics that impact on successful station efforts than do national sales. While spot analysts may not know to the dollar how spot sales will be affected by economic factors, the factors to look for are clear. In local markets, on the other hand, single industries can have an impact on station revenues. Predictability of local sales is contingent upon strong knowledge of the market.

Competitive factors are sometimes more predictable than local market conditions. There are two types of local competition that stations must monitor: competition from other stations and from other media. Unlike product-based industries, television stations are easy to monitor because their inventory is publicly aired. In addition to the amount and type of inventory, the pricing of inventory is important. Through Arbitron and Nielsen all television stations in a given market have access to comparative audience data. Television is in a unique position to monitor competitors very carefully and adjust strategies accordingly.

The station sales department must prepare a *rate card,* which lists the cost of every length of spot for every program, based on the Nielsen and Arbitron ratings, and on the demographics that offer a particular audience likely to be interested in the advertiser's particular product or service. (See Figure 3–4 for an example of a rate card that indicates availabilities and Figure 3–5 for an example of a card that indicates discounts for frequency of announcements.)

Rates are not always the same for all advertisers. For example, a rate card may have *fixed, semi-fixed,* and *preemptible* rates. The latter, for example, offers a lower rate but provides for preemption of the commercial if a higher rate buyer wants that spot, with the preempted commercial rescheduled in a time slot as close as possible to the original. Different rates apply, as well, to *rotations,* which are of two types: *horizontal* (the ad is carried at exactly the same time Monday

MARKET: BINGHAMTON

STATION: **WICZ-TV 40**

STATION AVAILABILITIES: PRIME

60" - DOUBLE 30" RATE

10" - 60% 30" RATE

DATE: 9/1/87

PAGE:

DAY/ TIME	PROGRAM	10"	MAX. 30" P/W	30" AVAILABLE WEEK OF: 9/7	9/14	9/21	9/28			RATE QTR. 3	QTR. 4
Mon. 8-9PM	ALF/Valerie		7	OPEN	OPEN	OPEN	OPEN			200	225
Mon. 9-11PM	Monday Movie		8	OPEN	OPEN	OPEN	OPEN			225	275
Tues. 8-9PM	Matlock		7	OPEN	OPEN	OPEN	OPEN			250	275
Tues. 9-11PM	NBC Movie		8	OPEN	OPEN	OPEN	OPEN			250	275
Wed. 8-9PM	Highway to Heaven		7	OPEN	OPEN	OPEN	OPEN			325	350
Wed. 9-10PM	Bronx Zoo		4	OPEN	OPEN	OPEN	OPEN			225	275
Wed. 10-11PM	St. Elsewhere		6	OPEN	OPEN	OPEN	OPEN			225	250
Thurs. 8-9PM	Cosby/ Various		7	TIGHT	TIGHT	TIGHT	TIGHT			850	950
Thurs. 9-10PM	Cheers/ Night Court		4	TIGHT	TIGHT	TIGHT	TIGHT			700	800
Thurs. 10-11PM	L.A. Law		6	TIGHT	TIGHT	TIGHT	TIGHT			350	450
Fri. 8-9PM	Rags to Riches		6	OPEN	OPEN	OPEN	OPEN			175	200
Fri. 9-10PM	Miami Vice		4	OPEN	OPEN	OPEN	OPEN			500	550
Fri. 10-11PM	Crime Story		6	OPEN	OPEN	OPEN	OPEN			300	325
Sat. 8-9PM	Fact/227		8	OPEN	OPEN	OPEN	OPEN			200	225
Sat. 9-10PM	Golden Girls/Amen		4	OPEN	OPEN	OPEN	OPEN			350	400
Sat. 10-11PM	Hunter		6	OPEN	OPEN	OPEN	OPEN			150	200
Sun. 7-8PM	Our House		4	OPEN	OPEN	OPEN	OPEN			100	125
Sun. 8-9PM	Family Ties		6	OPEN	OPEN	OPEN	OPEN			175	400
Sun. 9-11PM	Sunday Movie		10	OPEN	OPEN	OPEN	OPEN			250	275
Tues-Fri 9PM	Prime Time Newsbreak			OPEN	OPEN	OPEN	OPEN			300	350
Mon-Sun 10PM	Prime Time Newsbreak			OPEN	OPEN	OPEN	OPEN			300	350

WICZ-TV 40 P.O. BOX 1626, BINGHAMTON, NEW YORK 13902 – (607) 770-4040 / TWX NUMBER: (510) 252-1998

MARKET: BINGHAMTON

STATION: **WICZ-TV 40**

STATION AVAILABILITIES: MONDAY-FRIDAY

60" - DOUBLE 30" RATE

10" - 60% 30" RATE

DATE: 9/1/87

PAGE:

DAY/ TIME	PROGRAM	10"	MAX. 30" P/W	30" AVAILABLE WEEK OF: 9/7	9/14	9/21	9/28			RATE QTR. 3	QTR. 4
6-6:20AM	Jimmy Swaggart		10	OPEN	OPEN	OPEN	OPEN			15	15
6:30-7AM	NBC News at Sunrise	OK	15	OPEN	OPEN	OPEN	OPEN			15	20
7 - 9AM	Today Show	OK	120	OPEN	OPEN	OPEN	OPEN			30	40
9-10AM	PTL Club		10	OPEN	OPEN	OPEN	OPEN			15	15
10AM-12N	A. M. Rotation	OK	60	OPEN	OPEN	OPEN	OPEN			15	20
12N-3PM	P. M. Rotation	OK	110	OPEN	OPEN	OPEN	OPEN			15	20
3 - 5PM	Kids		150	TIGHT	TIGHT	TIGHT	TIGHT			40	50
5-5:30PM	Mork and Mindy (Eff. 9/14 Diff. Strokes)		55	OPEN	OPEN	OPEN	OPEN			50	80
5:30-6PM	Hollywood Squares (9/14 & 9/21 Mork & Mindy)	OK	55	OPEN	OPEN	OPEN	OPEN			80	125
6-6:30PM	Local News	OK	160	OPEN	OPEN	OPEN	OPEN			20	25
6:30-7PM	Network News		10	OPEN	OPEN	OPEN	OPEN			25	35
7-7:30PM	Gimme a Break		80	OPEN	OPEN	OPEN	OPEN			100	250
								Rotation		90	200
7:30-8PM	WKRP in Cincinatti		75	OPEN	OPEN	OPEN	OPEN			100	250
11-11:30PM	INN: USA Tonight		30	OPEN	OPEN	OPEN	OPEN			20	25
11:30PM-12:30AM	Tonight Show		75	OPEN	OPEN	OPEN	OPEN			25	35
12:30-1:30AM	Letterman		48	OPEN	OPEN	OPEN	OPEN			15	20

WICZ-TV 40 P.O. BOX 1626, BINGHAMTON, NEW YORK 13902 – (607) 770-4040 / TWX NUMBER: (510) 252-1998

STATION AVAILABILITIES: SATURDAY & SUNDAY

MARKET: BINGHAMTON
STATION: WICZ-TV 40

60" - DOUBLE 30" RATE
10" - 60% 30" RATE

DATE: 9/1/87
PAGE:

DAY/ TIME	PROGRAM	10"	MAX. 30" P/W	30" AVAILABLE WEEK OF: 9/7	9/14	9/21	9/28			RATE QTR 3	QTR 4
SATURDAY											
7:30-11AM	Kids Rotator	OK	32	TIGHT	TIGHT	TIGHT	TIGHT			30	35
11AM-12NN	Kids		8	OPEN	OPEN	OPEN	OPEN			30	35
12NN- 5:30PM	Sports/Various		30	OPEN	OPEN	OPEN	OPEN			75	80
								College Football		125	
5:30-6PM	Siskel & Ebert		28	OPEN	OPEN	OPEN	OPEN			25	30
6 - 8PM	The Big Movie	OK	20	OPEN	OPEN	OPEN	OPEN			75	80
11-11:30PM	Throb		12	OPEN	OPEN	OPEN	OPEN			30	30
11:30PM-1AM	Saturday Night Live		28	OPEN	OPEN	OPEN	OPEN			25	30
SUNDAY											
7:30-11AM	Religion		22	OPEN	OPEN	OPEN	OPEN			20	20
11AM-12:30p	Various		7	OPEN	OPEN	OPEN	OPEN			20	20
12:30-1PM	Meet the Press		4	OPEN	OPEN	OPEN	OPEN			30	30
1 - 5:30PM	Sports/Various		16	OPEN	OPEN	OPEN	OPEN			75	
								Pro Football		175	
5:30 - 6PM	Various		4	OPEN	OPEN	OPEN	OPEN			75	
6 - 6:30PM	Community Scene		4	OPEN	OPEN	OPEN	OPEN			20	20
6:30 - 7PM	Network News		4	OPEN	OPEN	OPEN	OPEN			30	30
11-11:30PM	What a Country		6	OPEN	OPEN	OPEN	OPEN			30	30
11:30PM-12:30AM	PTL Club		2	OPEN	OPEN	OPEN	OPEN			20	20
12:35-1:05AM	Siskel & Ebert		4	OPEN	OPEN	OPEN	OPEN			20	25

WICZ-TV 40, P.O. BOX 1626, BINGHAMTON, NEW YORK 13902 – (607) 770-4040 / TWX NUMBER: (510) 252-1998

Figure 3–4 Sample station ad availabilities for prime time *(left top),* Monday through Friday nonprime time *(left bottom),* and Saturday and Sunday nonprime time *(above).* (Courtesy WICZ-TV 40, Binghamton, NY.)

through Friday) and *vertical* (the ad appears at a different time each day).

Local time sales are made to the leading stores, services, and industries in the community. Usually, these local advertisers are virtually the same throughout the country: department stores, automobile distributors, supermarkets, restaurants (especially fast-food chains), banks, clothing, home furnishings, and, depending on the size of the community, entertainment such as movies, theatres, night clubs, athletic events, and pop and classical music. Of course, locales vary, and while one is not likely to find ads for farm equipment distributors on urban northeast stations, they are a staple on stations in many other parts of the country.

Spot costs for movies, specials, and sports events are illustrated in Figure 3–6.

While, as noted earlier, large advertisers use ad agencies, most local advertisers use their own staffs or the TV station's staff for developing their commercials. This means that the station's salesperson may have to work closely with the advertiser and serve as both an advertising and creative production expert. The salesperson, therefore, must have a thorough knowledge of the advertiser's business, as well as of the specific product or service being advertised, the relationship of the item to those of competitors, the actual and potential customers, the location and facilities of the advertiser, and any special problems or attributes that should be considered in preparing the ad. When it's time for production, the station's sales staff must have a highly cooperative working relationship with the station's production de-

40
WICZ·TV

WICZ-TV
9/87

MONDAY THROUGH SUNDAY PRIME TIME

30 Second

1	2	3	4	5	6	7	8	9	10
$1050	$950	$800	$700	$600	$500	$400	$300	$250	$200

SF Rate: 1500

Monday	8-9PM	NBC Network
	9-11PM	NBC Network
Tuesday	8-9PM	NBC Network
	9-10PM	NBC Network
	10-11PM	NBC Network
Wednesday	8-9PM	NBC Network
	9-10PM	NBC Network
	10-11PM	NBC Network
Thursday	8-9PM	NBC Network
	9-10PM	NBC Network
	10-11PM	NBC Network
Friday	8-9PM	NBC Network
	9-10PM	NBC Network
	10-11PM	NBC Network
Saturday	8-9PM	NBC Network
	9-10PM	NBC Network
	10-11PM	NBC Network
Sunday	7-8PM	NBC Network
	8-9PM	NBC Network
	9-11PM	NBC Network

Figure 3–5 TV ad rate cards corresponding to time slots in Figure 3–4. Rates are for prime time *(left)* and daytime and weekend nonprime time *(right)*. (Courtesy of WICZ-TV 40, Binghamton, NY.)

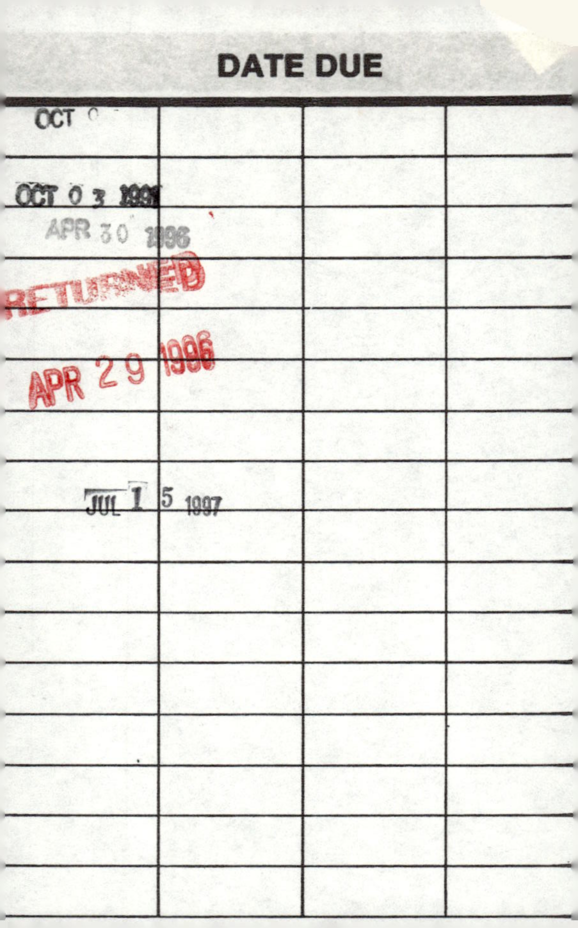
DATE DUE
OCT
APR 30 1996
RETURNED
APR 29 1996
JUL 15 1997

40 WICZ·TV

TV RATE CARD
SEPTEMBER, 1987

WICZ-TV
BINGHAMTON, N Y
N B C

F = Fixed
I = IMMEDIATELY PRE-EMPTIBLE

30 Seconds

DAYTIME & FRINGE			1	2	3	4	5	6
Mon-Fri	6:30-7AM	News at Sunrise	$50	$35	$30	$25	$20	$15
Mon-Fri	7-9AM	Today Show	75	45	40	35	30	25
Mon-Fri	10A-12NN	NBC AM Rotation	50	35	30	25	20	15
Mon-Fri	12N-3PM	NBC PM Rotation	50	35	30	25	20	15
Mon-Fri	3-330PM	Smurfs	75	65	55	50	45	40
Mon-Fri	330-4PM	Dennis the Menace	75	65	55	50	45	40
Mon-Fri	3-5PM	Kids' Rotator	75	65	55	50	45	40
Mon-Fri	4-430PM	Duck Tales	75	65	55	50	45	40
Mon-Fri	430-5PM	ThunderCats	75	65	55	50	45	40
Mon-Fri	5-530PM	Diff'rent Strokes	100	85	75	60	50	40
Mon-Fri	530-6PM	Family Ties	175	125	100	90	75	60
Mon-Fri	6-630PM	Early News	75	45	30	25	20	15
Mon-Fri	630-7PM	NBC News	85	45	40	35	30	25
Mon-Fri	7-730PM	M*A*S*H	300	250	200	175	150	125
Mon-Fri	730-8PM	Cheers	300	250	200	175	150	125
Mom-Fri	11-1130PM	INN: USA Tonight	75	45	30	20	15	10
Mon-Fri	1130P-1230A	Tonight Show	75	45	40	35	30	25
Mon-Fri	1230-130AM	David Letterman	50	35	30	25	20	15
WEEKEND								
Saturday	730AM-12NN	NBC Kids' Rotation	50	45	40	35	30	25
Saturday	12NN-5PM	NBC Sports		See Sports and Specials				
Saturday	530-6PM	Siskel & Ebert	90	75	65	60	55	45
Saturday	6-8PM	The Big Movie	125	100	80	70	60	55
Saturday	11-1130PM	Sat. Night Home Show.	55	40	35	30	25	20
Saturday	1130PM-1AM	Saturday Night Live	55	40	35	30	25	20
Sunday	730-11AM	Religious/Various	50	40	35	30	25	20
Sunday	11AM-12NN	Various	50	35	30	25	20	15
Sunday	12NN-5PM	Various	75	40	35	30	25	20
Sunday	1-7PM	NBC Sports		See Sports and Specials				
Sunday	11-1130PM	Throb	75	40	35	30	25	20
Sunday	1235-105AM	Siskel & Ebert	70	50	40	35	25	20

Announcements Between Time Blocks take the rate of the higher Time Block

60 Seconds = Double the 30 Seconds rate.
10 Seconds = 60% of the 30 Seconds rate rounded to the next dollar.
15 Seconds = Will accept 75% of the 30 Seconds rate rounded to the next dollar

Rate Card Provisions- Station subscribes to the following coded provisions in SRDS: 1a, 2a, 3a, 4a, 5, 6a, 10i, 11m, 12m, 13m, 14c, 20a, 40a, 40b, 41a, 61a, 61b, 70h, 82, 84, 86, 87b, 87c.

Issued 9/87

40 WICZ-TV
BINGHAMTON N.Y.

memo

FROM: PROGRAMMING
TO: SALES

NBC MOVIES & SPECIALS SEPT., OCT., 1987

Day	Date	Time	Program	Cost
Sun.	9/6	7-8PM	OUR HOUSE (Conc.)	$125
		9-11PM	NBC MOVIE: "Coast to Coast"	$275
Mon.	9/7	9-11PM	NBC MOVIE: "A Year in the Life" (Conc.)	$275
Tues.	9/8	9-11PM	NBC MOVIE: "Beyond the Limit"	$275
Thurs.	9/10	9:30-10PM	NIGHT COURT (Conc.)	$800
		10-11PM	L.A. LAW (Part 2)	$450
Fri.	9/11	8-9PM	ALF LOVES A MYSTERY/AMAZING STORIES	$200
Sun.	9/13	7-8PM	OUR HOUSE (Premiere)	$125
		8-8:30PM	FAMILY TIES (Premiere)	$400
		9-11PM	NBC MOVIE: "Private Eye: (Preview)	$275
Mon.	9/14	9-11PM	NBC MOVIE: "Irreconicalable Differences"	$275
Tues.	9/15	9-11PM	NBC MOVIE: "Killer in the Mirror"	$275
Wed.	9/16	8-9PM	HIGHWAY TO HEAVEN (Part 1- Premiere)	$350
		9-10PM	A YEAR IN THE LIFE (Premiere)	$275
		10-11PM	ST. ELSEWHERE (Premiere)	$250
Thurs.	9/17	8:30-9:30PM	NBC INVESTIGATES BOB HOPE	$950
		9:30-10PM	NIGHT COURT (Premiere)	$800
		10-11PM	L.A. LAW (Conc.)	$450
Fri.	9/18	8-10PM	RAGS TO RICHES (Premiere)	$200
		10-11PM	PRIVATE EYE (Premiere)	$325
Sat.	9/19	9-9:30PM	GOLDEN GIRLS (Premiere)	$400
		9:30-10PM	MAMA'S BOY (Premiere)	$400
		10PM-12M	1987 MISS AMERICA PAGEANT	$400
Sun.	9/20	7-8PM	OUR HOUSE (Conc.)	$125
		8:30-9:30PM	MY TWO DADS (Premiere)	$400
		9-11PM	NBC MOVIE: "Highwayman"	$275
Mon.	9/21	8-9PM	ALF/VALERIE'S FAMILY (Premiere)	$225
		9-11PM	NBC MOVIE: "If It's Tuesday, It Must Be Belgium"	$275
Tues.	9/22	8-10PM	MATLOCK (Premiere)	$275
		10-11PM	CRIME STORY (Premiere)	$275
Wed.	9/23	8-9PM	HIGHWAY TO HEAVEN (Conc.)	$350
Thurs.	9/24	8-9PM	THE COSBY SHOW/A DIFFERENT WORLD (Premiere)	$950
		9-10PM	CHEERS (Premiere)/NIGHT COURT (Conc.)	$800
Fri.	9/25	9-10PM	MIAMI VICE (Premiere)	$500
Sat.	9/26	8-9PM	FACTS OF LIFE P-1; Premiere)/227 (Premiere)	$200
		9:30-10PM	J.J. STARBUCK (Preview)	$350
Sun.	9/27	9-11PM	NBC MOVIE: "The Terminator"	$275
Mon.	9/28	9-11PM	NBC MOVIE: "Assault & Matrimony"	$275

Figure 3–6 Sample spot costs for movies and specials *(left)* and sports events *(right).* (Courtesy of WICZ-TV 40, Binghamton, NY.)

memo

FROM: PROGRAMMING
TO: SALES

NBC SPORTS, SPORTSWORLD — SEPT., OCT., 1987

Day	Date	Time	Program	Price
Sun.	9/6	2-6PM	1987 WORLD CHAMPION.TRACK & FIELD	$75
Sat.	9/12	2-5PM	MAJOR LEAGUE BASEBALL	$100
Sun.	9/13	12:30-7PM	NFL FOOTBALL DOUBLEHEADER	$175
Sat.	9/19	2-5PM	MAJOR LEAGUE BASEBALL	$100
Sun.	9/20	12:30-4PM	NFL FOOTBALL	$175
Sat.	9/26	Time Tent.	MAJOR LEAGUE BASEBALL: Wild Card	$400
Sun.	9/27	12:30-7PM	NFL FOOTBALL DOUBLEHEADER	$175
Sat.	10/3	Time Tent.	MAJOR LEAGUE BASEBALL: Wild Card	$400
Sun.	10/4	12:30-4PM	NFL FOOTBALL	$175
Tues.	10/6	8-11PM	BASEBALL: NAT'L. LEAGUE EAST: Game 1	$400
Wed.	10/7	3PM-CC	BASEBALL: NAT'L. LEAGUE EAST: Game 2	$400
		8-11PM	BASEBALL: AM LEAGUE WEST: Game 1	$400
Thurs.	10/8	8:30PM-CC	BASEBALL: AM LEAGUE WEST: Game 2	$400
Fri.	10/9	8-11PM	BASEBALL: NAT'L. LEAGUE WEST: Game 3	$400
Sat.	10/10	1PM-CC	BASEBALL: AM. LEAGUE EAST: Game 3	$400
		4-6PM	SPORTSWORLD	$75
		8-11PM	BASEBALL: NAT'L LEAGUE WEST: Game 4	$400
Sun.	10/11	12:30-4PM	NFL FOOTBALL	$175
		4:30PM-CC	BASEBALL: AM. LEAGUE WEST: Game 5	$400
		8-11PM	BASEBALL: AM LEAGUE EAST: Game 4	$400
Mon.	10/12	3PM-CC	BASEBALL: AM. LEAGUE EAST: Game 5	$400
Tues.	10/13	8-11PM	BASEBALL: NAT'L LEAGUE EAST: Game 6	$400
Wed.	10/14	3PM-CC	BASEBALL: AM. LEAGUE WEST: Game 6	$400
		8-11PM	BASEBALL: NAT'L LEAGUE EAST: Game 7	$400
Thurs.	10/15	8:30-11PM	BASEBALL: AM. LEAGUE WEST: Game 7	$400
Sun.	10/18	12:30-7PM	NFL FOOTBALL DOUBLEHEADER	$175
Sun.	10/25	12:30-7PM	NFL FOOTBALL DOUBLEHEADER	$175
Sat.	10/31	2-3:30PM	PBA FALL TOUR	$100
		3:30-5:30PM	SPORTSWORLD	$75

partment inasmuch as the production department must be brought in on the completion of the planning and take over the actual production of the commercial.

There is an important advantage to clients who advertise locally. Cost factors make it possible sometimes to sponsor an entire program and to develop name recognition in connection with that program, rather than to be just another spot ad on a show. With local, nonaffiliated stations, mostly UHFs, local advertising is of special significance; such stations do not have national programs that can attract national spots.

Deregulation

Deregulation and the demise of the National Association of Broadcasters' Codes are perceived by many as a boon to station management, but are seen by others as harmful to sales. With the increasing costs for commercials, advertisers have had to advertise less frequently. Over the years, for example, program sponsors (that is, firms that sponsored an entire program) became fewer and fewer, and more and more advertisers became alternate sponsors (alternating weekly with other compatible sponsors), multiple sponsors (where a number of sponsors shared the costs for a program), coop ads (where the manufacturer and the retailer split the costs), and finally, a most common form today, the placement of one or two commercials in a program that has many such individual ads. The latter is especially prevalent on high-rated programs where the per-insertion cost is exceptionally high.

With no FCC rules or industry-wide codes any longer affecting commercial time, a sponsor may turn out to be just one of a plethora of advertisers in upwards of twenty minutes of commercial time on a one-hour program. In fact, when the FCC deregulated its commercial time limits and its program-length commercials rule (which essentially discouraged advertising that continued for more than five straight minutes), television was able to insert as many commercials as the traffic would bear. A sponsor now has to pick and choose carefully to be certain not only that the ad reaches the right target audience in the right market at a reasonable cost per thousand people reached (CPM), but that the ad doesn't get lost in a morass of other commercials in the same time segment.

THE SALES STAFF

Director of Sales or Sales Manager

The *director of sales* or *sales manager* (or in an increasing number of stations the *vice president for sales*) is responsible for the success of the entire department and, by projection, for the financial solvency of the station.

It is critical that the director hire the best people for the other sales staff jobs. The director works with the local sales manager to motivate salespersons and to ensure that the right salespersons are matched to the right accounts. To sustain morale the director may even become involved in adjusting accounts so that all salespersons have a more or less equal chance at commissions. On the national level, the sales director works with the national sales manager, who maintains close and continuous contact with the station rep.

Another critical function of the sales director is to set the departmental budget. The projected revenues for this department become the projected revenues for the station. Part of revenue projection involves setting advertising rates. The rate-making function is dependent on having control of the station's time inventory. Although the director of sales does not carry an account list, except in some of the smaller stations with limited staff, the director does represent the department as a whole to outside organizations and frequently is involved directly in negotiations with major clients—for example, a franchise to carry the games of an athletic team. When dealing with outside corporate entities, the director of sales represents and reports to the station manager.

The director of sales must have an understanding not only of the station, the television industry, and the particular market, but of the competition. The competition is defined as not just other television stations, but is all advertising outlets that are competing for a slice of the entire advertising pie in the market: newspapers, magazines, direct mail, billboards, subway cars, buses, and radio. The sales manager, with the general manager, constantly evaluates the extent of that advertising pie, how much of it is going for television in general, and how much of that their station might get. Wise management determines not only what has to be done to increase its portion. When the bottom may be too low under any circumstance, no matter what the sales manager does, because the pie itself is just too small, management must decide whether to cut back or to sell the station.

The sales director should have a background in retailing, preferably not only in the media, but also in one or more of the major advertisers' fields. Generally, a director of sales has held the positions of national and local sales manager before being promoted. He or she must know market research, including the functions of specific ratings companies, and must be able to initiate, supervise, and evaluate local demographic studies.

The ability to supervise and maintain efficient records on all members of the sales staff is essential. The director has to determine which individuals need help and which general areas of sales are weak or are increasing. Prompt action is critical: problems cannot be ignored or they may compound beyond solution. Periodic conferences with staff and individual salespersons are necessary. These conferences sometimes take place once a day in addition to special meetings of the entire staff regarding new programming, station promotions, new products, new potential advertisers, and similar matters.

The director of sales is paid on a salary basis, and some stations will pay a bonus, or override, on exceptional sales records of the department. Specific salaries vary according to the size of the market.

National Sales Manager

The *national sales manager* is primarily responsible for dealing with the station rep firm. Because the firm represents the station to national and regional advertising agencies, the national sales manager frequently is involved in negotiating major client deals. Qualifications for this position include experience in research, knowledge of the national advertising market, ability to manage inventory, and, of course, strong interpersonal skills. This position is paid solely on salary, in most cases.

Local Sales Manager

The *local sales manager* is responsible for motivating and supervising the salespersons or account executives and is usually the one who draws up the account lists for the salespersons. In difficult client negotiations the local sales manager may step in. This requires strong knowledge of the local market. Motivating the staff includes meeting frequently on the status of the station and its various departments, especially those related to the sales function, helping to find leads for new business, and informing individual salespersons of new techniques for selling and of any feedback on their current approaches and techniques. Some local sales directors carry their own account list of a few, select clients. Generally, this position is a salary-only position, with some stations offering optional bonuses or overrides on local sales.

Salespersons or Account Executives

At the heart of the sales department are the *salespersons* or *account executives*. An account executive's job entails more than making a sale. Accounts must be serviced after the sale. Salespersons work with clients to arrive at the best combination of spots, and with the station's producers to make sure that the production of a spot is both creative and effective. The salesperson must know the market in order to best target the advertiser's selling efforts. He or she must also understand the implications of quantitative rating data. Good skills in oral and written communication are essential, as is a sense of creativity. Sales is a "people" job and a salesperson must be friendly and outgoing and be able to convey a sincere interest in the advertiser's business and product or service. Aggressiveness must be translated into persuasiveness. And successful salespersons have to have sturdy egos. They must be able to bounce back quickly from rejection, even in pursuing the same client who rejected them.

In a classic study of what makes a good salesperson, David Meyer and Herbert Greenberg identified empathy and ego drive as key components of effectiveness.[10] The salesperson must empathize with the client. The salesperson must have a strong ego drive, a need to make the sale to fulfill a personal need beyond that of monetary income. If both empathy and ego are not present, the salesperson likely will not close the sale or will be perceived as too aggressive, limiting future success with the client.

In achieving empathy the salesperson can apply a number of proven techniques: have an intimate knowledge of or past experience with that particular product or service; make friends with the advertiser, get into warm conversational situations on a personal as well as professional level; offer that little bit of extra attention and care about the advertiser's business and product or service that the competitive television stations and media may not provide. The competition in most instances is still the printed ad, and the salesperson must be able to show how the visual ad can be more effective, at a low CPM. Although the best salesperson appears to know the market, know merchandising, and know the language of sales and advertising when initially meeting the prospective client, inexperienced persons are not to be ruled out. A station always has to have fresh newcomers and finding people with drive and personality provides the station with an opportunity to teach them the information and techniques and mold their work in the station's image.

It is important that salepersons maintain control over their own accounts. For example, no one in programming should contact the salesperson's account regarding a particular ad or about a program the ad will be on without first clearing it with that salesperson. The management should support the work of the salesperson by impressing on the rest of the station's departments and staff that without the salesperson's successful work, they would all be out of jobs. For the salesperson, homework is critical. Long and flexible hours are essential. Frequently, one's personal life must be preempted. Before approaching a client the salesperson must know everything about the market, the product, and the demographics so that the time and energies of all parties involved are most efficiently used.

Jennifer McCann, sales manager for radio station WJIB in Boston, has prepared a list of dos and don'ts on radio time selling that can be interpolated, as follows, for television:[11]

The relationships salespersons establish with clients are as important as rate cards. Cost-per-thousand and time slots mean little if the advertiser does not develop a trust in the account executive and in the station.

The salesperson should always be prepared for all eventualities when meeting a client. This means determining all possible objections beforehand and preparing materials that will answer those concerns. If a question arises that the salesperson cannot answer, faking a response is not the solution. The salesperson should simply tell the client that after getting the answer, he or she will call back. The salesperson should not forget to do it immediately.

Time is a precious commodity, to the client as well as to the salesperson. Always being on time and not making excuses is an essential trait. If a salesperson is going to be unavoidably late for a meeting with a client, making certain that the client is informed as soon as possible saves the client from wasted waiting time. To avoid embarrassing a client who forgot to make a notation in the appointment book or who forgets to call when a more critical matter comes up, always phoning ahead to reconfirm the appointment and time is a good practice.

A hard sell rarely works, especially on clients who have done a fair amount of advertising. A salesperson must be polite, even with clients who are not, and listen to what they have to say. Clients seek advice from account executives and if concern for the client's needs is conveyed, the client is more likely to take the advice to buy time on that salesperson's station. Being upbeat and creating a feeling of success about one's station, as well as taking a positive approach by showing what the station can do for the client, rather than trying to tear down competing stations, is a more professional and pleasant way of doing business.

Even if the client does not buy, the salesperson should thank the client for considering the station. The client may be more willing to buy next time. The salesperson should try to find out why the client didn't buy in order to create a more effective sales approach to that client next time. Inquiring about a client's advertising in general and finding out how that client's product or service does in different media buys is research worth doing.

A salesperson should, above all, follow through. After making a sales pitch a salesperson cannot wait for the client to offer to buy. The salesperson has to ask. And if the circumstances are such that it does not appear to be high-pressure, after a client's refusal and further discussion it is a good idea to ask again for an order.

As a salesperson plans a campaign to sell to a specific client, he or she must remember that the client may not care about the station or even about television. The client's interest is in selling a product or service. Therefore, a salesperson needs to be thoroughly familiar with the client's product, and be able to talk about it with the same excitement and commitment that the client has. That sometimes takes a great deal of research: history of the product or service, its impact and acceptance in the service area, its substance (which may involve interviewing experts in the field), its competition, its economic ups and downs, its potential customers (demographics), and similar factors. A salesperson may spend more time in the library with periodicals and newspapers in developing a sales bid than when writing a term paper on the subject in college.

Time is saved by determining who is the right person to see at the client's firm. A salesperson will do well to find out whether the firm is in an expansion mode and ready to buy new or additional time. While it may be necessary to spend time with someone lower down on the corporate ladder to get to the person who can authorize the purchase of advertising time, one must remember where the important sales pitch has to be made.

To clients new to the electronic media field, selling the idea of television as a successful sales tool may have to come before selling one's own station as the specific way to go.

It is important to remember that although the client may look to an account executive for advice (while being aware that that salesperson has a very obvious vested interest), the client does not appreciate being told what to do. By establishing oneself as a member of the client's sales team, a salesperson can more easily determine the best way to go; a well prepared salesperson will make that way his or her station. As the expert on television advertising, the salesperson can be the one with new ideas that will make the client's business successful. By the same token, it is important to remain open to new ideas from the client, even if they do not fit the past practices of one's station.

Above all, a salesperson must keep trying. It sometimes takes six or more calls on a client to sell time. It helps to find legitimate reasons to stop and see the client occasionally. On some of these occasions a salesperson will find the opportunity to ask again for an order. If using a sales pitch, a salesperson should try to incorporate new approaches rather than using the same one used before. Good client rapport and perseverance usually result in success.

Salespersons or account executives work mostly on commission. Compensation plans vary from station to station, but approximately 60–70% of what salespersons earn is from commission on sales; the rest is guaranteed salary. At some stations salespersons are given greater compensation for bringing in new business or for making direct sales to clients than for sales made through advertising agencies. The larger the market, the higher the cost of advertising time and, in turn, the higher the salaries. Not only are salaries different in large versus small markets, but the day-to-day earnings and even positions of salespersons may vary as well.

Beyond sales, the salesperson represents the station in another way: he or she is the primary and most direct live contact between the station and the local business community. A large part of the station's image can be shaped by the sales staff.

PROMOTION

In a competitive market, effective promotion is essential if the sales department is to make the product attractive to the advertiser. While on the surface the advertiser invests in the station and its programs, the product in fact is the size and composition of the audience the station delivers. While the promotion department literally promotes the programs and the station's image to lure more viewers, the end product is the number and demographics of these viewers. In this respect, promotion (and advertising) has become more and more marketing oriented, closely tied to the aims of the sales department.

Large stations usually have a separate promotion department (see Chapter 2). Smaller stations may put the promotion office or person under the general manager's office, in programming, or in sales and advertising. Inasmuch as the job of the promotion department is literally to promote the station's programs and other activities for the purpose of attracting sponsors to the station image and/or to specific programs, its work must be coordinated carefully with sales and advertising, as well as with programming and overall managerial goals.

As noted in Chapter 2, in addition to creating an image for the station, promotion includes publicizing ongoing programs, creating interest in locally produced shows, including the news, and doing "tease" advertising for specific shows. Promotion frequently works with its network (if affiliated) or group owner (if it has one) for mutual publicity. The promotion department may sometimes become a production unit, as well, creating spots for airing and tapes for distribution in the community. Its work is not limited to video, however, and may include advertising in other media, such as print and billboards, and may go beyond advertising as such to include contests and other devices that will attract attention to the station.

Promotion frequently is involved directly with ratings information, and uses it to determine the demographics for its outreach. By the same token, promotion steps up its operations for rating period *sweeps*.

As noted earlier, promotion not only must relate the station effectively to the public, but to potential sponsors, as well. On the national level, in publications such as *Broadcasting,* for example, and on local levels, on buses and billboards, for example, materials promoting the station stress such factors as audience size and composition, the place of the station in the market, station awards, and good station ratings in order to catch, primarily, the attention of the potential advertiser and, secondarily, that of the viewer.

Inside Promotion

The logical means for promoting one's own programs are on one's own station. (It is not likely, of course, that one will be able to buy air time on a competitor's station!) On-air spots vary from a brief announcement of an upcoming program to an elaborate commercial with excerpts from the show, designed to whet the viewer's appetite. Placement of a *promo* can vary from a voice-over announcement during the ending credits for a show, to a graphic shown along with the station ID during a break, to a thirty- or sixty-second spot in a place where no time has been sold. One should be aware, however, that the sales people can preempt a promo if they come up with a paid commercial for that slot.

The sponsor frequently is part of on-air promotion. For example, discount coupons, special premiums, treasure hunts, and other techniques may relate to the sponsor's product or service or place of business. Promoting the station itself to enhance ratings may also involve incorporating on-air approaches, such as clues for contests or call-ins to specified shows (locally produced, of course) for prizes.

Outside Promotion

Regular press releases must be a part of every promotion department's work. Schedules of the station's programs must be made available daily to the press, and should have accurate up-to-date information on the title, date, and time of showing, key performers, especially stars, special guests (or hosts if it's not a drama), a summary of the show and, hopefully, something unique, even a justifiable gimmick, that makes the program stand out.

Press releases on individual programs are important, especially when they include something startling (an exclusive news interview, a revelation in the private life of a star, a human interest connection to the role the performer is playing).

Almost all newspapers have a television section that not only includes schedules and stories, but highlights of the day's offerings. The promotion department submits the station's schedule with annotated program descriptions that will have special appeal to the area's viewers.

Purchasing space in another medium is also an option, depending on the promotion budget of the station. Newspapers and radio stations are obvious choices. Depending on frequency of publication and whether the station is promoting a single program or a series, or seeking institutional advertising for the station's image, magazines are another outlet. Posters, billboards, bumper stickers, skywriting, tickets for ball games, testimonials from athletes, fund raising for worthy causes, and star benefits are all possible promotion techniques.

Some programs lend themselves to serious, in-depth promotion. For example, docudramas, documentaries, and educational programs are ready made for viewer guides that can be distributed among civic groups and in schools.

Promotion must use exciting techniques, but cannot be based on gimmicks alone. Basic to any determination of what and how to promote is an understanding of the actual and potential audience, the advertisers and their products or services, and the promotion being used by the station's competitors. This requires close cooperation with whoever is doing demographic research for the station, and whoever is interpreting the ratings.

Nonaudience Promotion

Promotion is not only promoting the station and its programs to viewers. Under the rubric of terms such as marketing, public relations, publicity, merchandising, and sales promotion one should apply to advertisers some of the same techniques used to promote the station to viewers. Sales promotions include strong institutional advertising, discounts on time buys, trial spots, and time and day (daypart) preferences, among others.

The Promotion Director

The promotion director must know the demographics of the viewers and potential viewers and the special needs and interests of the advertisers and potential advertisers. The promotion director must work closely with the station manager in determining the optimum as well as the pragmatic promotion budget for the station. In large markets a promotion budget for a single TV station can run into millions of dollars a year.

Part of the promotion director's responsibility, as is that of the station manager and sales director, is to become a visible part of the community. That requires, therefore, not only a knowledge of public relations and advertising, as well as television, but a personality conducive to socializing successfully with all segments of a community. It also takes a person who likes or at least can cope with being away from home almost every evening of the week and frequently on weekends to attend civic events, organizational meetings, educational activities, arts performances, and similar affairs.

Media consultant Maggie Dugan, a former station promotion coordinator and marketing director, counsels that the special abilities needed in a promotion director vary from station to station. The ideal promotion director, she says, "is creative, imaginative, enthusiastic, energetic, innovative, collaborative, diplomatic, articulate, organized, detail oriented, thorough." And each station, she adds, has its "own descriptives to add to the list." She describes the managerial style of promotion directors as one that involves the participation of staff members and is not dictatorial. "Promotion directors spend a lot of their time managing people who aren't 'officially' on their staff. They indirectly manage disk jockeys and salespeople."[12]

The closer the relationship of the promotion department to the marketing function of the station, the more the promotion director must know about the approaches to and techniques of selling. The promotion director's participation in community activities should include merchants and other potential sponsors in the circle of people to get to know.

So important is the position and work of the promotion staff that several decades ago a national Broadcasters Promotion Association was established, which provides conferences, workshops, and seminars, as well as information exchanges, contacts, and moral support for its members.

THE BOTTOM LINE

In the last few years established stations faced competition from new video options in the marketplace. One example is the growth in new VHF independent television stations nationally. Even with the traditionally lower ratings that independents garner, these stations can have a negative impact on existing stations. First, they further fragment the television "pie" in the market, at least initially. Second, such stations must come into the market with significantly lower rates, which may force tougher rate negotiations in the market overall. New independents can, on the other hand, bring in new television business by opening the medium to advertisers who were previously priced out of the market.

Yet another emerging video option in some markets is cable. While cable has not yet attracted audience numbers that can compete effectively with broadcast television, its impact is increasing and is likely to be felt first at the national level. In five years in the late 1980s, cable advertising revenues jumped from $58 million to $767 million, based on figures distributed by the Television Bureau of Advertising. It remains to be seen whether cable will expand or fragment the television advertising pie.

Among all local media, newspapers traditionally have dominated the advertising market. In spite of contractions within the newspaper business, this medium can be expected to continue local market dominance. Binghamton, Buffalo, and Philadelphia all lost one newspaper in recent years; however, television sales directors in those markets noticed no differences in their revenues. Todd Wheeler, director of sales at WPVI in Philadelphia, says that television does a terrible job of selling against newspapers. During a recent newspaper strike in that city, television stations gained new advertisers who had been using print. Once the strike was over, those advertisers returned to print.[13] Newspaper advertisers are very loyal in spite of rising costs and declining or stable circulation. This competitive market will most likely remain a challenge to television for years to come.

Market conditions and other media competitors do not alone explain the level of station profitability. The quality or popularity of programming, the pricing of the inventory, and the aggressiveness of the sales staff all play a role. If the station is facing profitabiity problems that are internal, there are several solutions. First, if the programming is not attracting a large audience, it may be a poor quality program or program flow for a given daypart may be poor. Program changes should be implemented if one of these conditions exist. Some programs achieve low audience shares not because they are poor programs but because they appeal to a narrow demographic. The answer may be not to eliminate the program but to sell it based on the narrowness or quality of the demographics viewing. Second, cutting the cost of station time may be another solution. Time may be priced over the prevailing cost per point in the market and advertisers may not be purchasing for that reason. Finally, are the sales staff aggressive in pursuing new clients and increasing sales to existing clients? This is the sales manager's responsibility to assess, especially if ratings are high but inventory is left unsold.

In summary, national, local, and internal factors mix to produce an impact on station sales. It is incumbent upon the sales department to bring in the revenues to offset the other department's expenses. The greatest expenses a station incurs are salaries and programming; both have a direct impact on revenues. Without quality staff and programming, the sales department cannot perform its function. Sometimes, in planning the budget, sales directors must set goals to accommodate expenses that are fixed and expected to increase. This may mean salaries or programming, but increases in sales targets can also reflect the more mundane aspects of station operation, like increases in utility costs, equipment replacement, or insurance. How does a sales director or a station manager plan the budget to account for all these uncontrollable market and nonmarket conditions?

Grant Reusch, director of finance and administration at WNBC, New York, lists three steps in planning sales growth. First, there is a need to identify market strength. Second, management must project CPM and sellout rates in every daypart. Third, is there special programming on the horizon that could radically alter the households using television?[14] For instance, viewing goes up with special events like the World Series or the Super Bowl. Market strength projection has been discussed above. How does management project CPM and sellout rates?

Most markets have a prevailing CPM or

cost per household rating point. It is based on market size, market demographics, and the time of the year. Publications such as the *Media Market Guide* list these costs by market. Market demographics sometimes overtake market size in determining cost per household rating point. For example, it costs more to reach a household in the wealthy community of Palm Springs, market 183, than to reach a household in market 182, El Centro-Yuma.

Determining a sellout rate is the most difficult task a manager faces. A manager must look at program stability and attractiveness, as well as market conditions. In election years, sellout rates can be expected to increase because there are more advertisements in the marketplace.

Setting both the sales department and the station budget is a planning process involving all departments of the station, including programming and news because they develop the product to attract viewers, and promotion and engineering because they incur expenses to support the station. The sales department's pivotal role in acquiring revenue makes its budgeting and planning process critical to all other station operations.

NOTES

1. Turow, Joseph. *Media Industries*. New York: Longman, 1984, pp. 10–13.
2. Ibid., p. 62.
3. *Television/Radio Age,* January 5, 1987, p. 85.
4. Schrank, William. Senior Vice President, Radio and Television Research, Katz Communications. Telephone interview with K. Mahoney, August 19, 1986.
5. "Bates Rates CPM Performance." *Marketing and Media Decisions,* August 1986, p. 46.
6. Fratrick, Mark. "How Important is Local Advertising to Today's Television Station?" *Broadcast Financial Journal.* Conference, 1986, p. 14.
7. Butterfield Communications Group. *Barter Syndication: Four Perspectives: Stations, Advertisers, Syndicators, Networks.* New York: Station Representatives Association, n.d., p. 8. (See also Lorimar-Telepictures. *Barter Syndication: A White Paper*. New York: Lorimar-Telepictures, 1985.)
8. Back, George. Chief Executive Officer, All American Television. Telephone interview with K. Mahoney, August 8, 1986.
9. Poltrack, David. *Television Marketing*. New York: McGraw-Hill, 1983, pp. 147–148.
10. Mayer, David and Herbert Greenberg. "What Makes a Good Salesman." *Harvard Business Review,* July-August, 1964, pp. 119–121.
11. McCann, Jennifer N. "Management: Sales and Advertising," in Hilliard, Robert L., *Radio Broadcasting*. New York: Longman, 1985, pp. 102–103.
12. *Broadcasting,* March 21, 1988, p. 21.
13. Wheeler, Todd. Director of Sales, WPVI-TV, Philadelphia. Telephone interview with K. Mahoney, November 4, 1986.
14. Rustin, Dan. *Inside the Media*. New York: International Radio and Television Society, 1984, p. 50.

BIBLIOGRAPHY

Alreck, Pamela and Robert Settle. *The Survey Research Handbook*. Homewood, IL: Dow-Jones-Irwin, 1985.

Dominick, Joseph R. and James E. Fletcher. *Broadcasting Research Methods*. Boston: Allyn and Bacon, 1985.

Eastman, Susan T. and Robert A. Klein. *Strategies in Broadcasting and Cable Promotion*. Belmont, CA: Wadsworth, 1982.

Heighton, Elizabeth J. and Don R. Cunningham. *Advertising in Broadcast and Cable Media*. Belmont, CA: Wadsworth, 1984.

Klepner, Otto. *Advertising Procedure*. 8th ed. Englewood Cliffs, NJ: Prentice-Hall, 1983.

Poltrack, David. *Television Marketing*. New York: McGraw-Hill, 1983.

Rustin, Dan. *Inside the Media*. New York: International Radio and Television Society, 1984.

Turow, Joseph. *Media Industries*. New York: Longman, 1984.

Warner, Charles. *Broadcast and Cable Selling*. Belmont, CA: Wadsworth, 1986.

Wimmer, Roger D. and Joseph R. Dominick. *Mass Media Research*. 2d ed. Belmont, CA: Wadsworth, 1987.

4

Programming

Robert L. Hilliard

Without the program, there is no paycheck.

The program is the product the station sells to the viewer, and, as noted in Chapter 3, the viewer is the product the station sells to the advertiser. Without the program, the engineering and technical staffs have nothing to send out over the air. There would be nothing to promote and no audience to measure. Programming is central to all other operations of the station. Good programming—or, more accurately, programming that draws large audiences—is necessary to the survival and growth of the station.

Where do these programs come from? How are they acquired? What do they cost? How are they scheduled? What keeps them on the air? How are they produced? Answers to these and other questions are the daily considerations of the TV station program manager and staff.

RESEARCH, RATINGS, AND DEMOGRAPHICS

Some programmers tell you that they make their choices by a "gut" feeling. Comparing their successes to those of programmers who assiduously analyze every variable affecting program choice sometimes makes one wonder whether all progress on television may be only a matter of luck and guesswork. Nevertheless, most successful television programmers will tell you that, as in any other field, research, evaluation, and planning on as scientific a basis as possible, plus some good intuition, are necessary. That's why, despite complaints about the power and validity of rating systems, program directors carefully evaluate all data, attempting to combine the variables into a meaningful guide to program selection and scheduling.

Audience research usually is classified as *quantitative* or *qualitative*. Quantitative research provides objective information about the audience, such as the number watching a given program, their age and gender breakdown, their economic and educational status, and other information providing as clear a picture as possible of the kinds of people watching. Qualitative research is more difficult to obtain. Here one would try to determine the attitudes of the public toward the program and, especially, toward the products or services advertised on the program. Such information gives greater depth to planning for effective programs and commercials.

A long-time criticism of the traditional surveys—by electronic devices attached to the set, by telephone, and by diary or survey—is that none of these techniques accurately determine whether anyone is really watching, or for how long a period. The A.C. Nielsen *audimeters,* for example, measure whether a set is on, but not whether there is even a single person in the room watching it. In 1988 the *people meter* emerged, a device that (depending on the manufacturer) either requires the pressing of a button whenever someone leaves or enters the viewing room, or features a heat sensor that registers bodies within viewing distance of the set.

The interpretation of research is the key to its value, and different evaluators frequently arrive at different results from the same research. Too often the testing of a new program reveals that it will be a hit, only to have it flop when aired, and vice versa. Networks have more at stake than local stations and either employ a competent in-house research team or hire an outside independent research firm. The local station usually depends on the local Nielsen or Arbitron ratings to provide a quantitative comparison between its numbers of viewers and those of its competitors. For qualitative research, a good local research firm is needed. Such research is not inexpensive.

What are some of the demographics determined by the rating systems that the programmer must take into consideration? Below are key areas presented in a Nielsen Report on the national TV scene. Local data is similar, interpolating these areas into local concerns.

- Stations and channels receivable per TV household and the percent share of these households
- Total persons in TV households
- TV ownership in households
- Percent of households using television
- Average hours of household usage per day
- Hours of TV usage per week by household characteristics (size, income, cable, non-adults)
- Weekly viewing activity for women, men, teens, children
- Persons viewing prime time
- Audience composition by selected program type of regularly scheduled network programs in prime time
- Video cassette recorder activity in VCR homes
- Prime time VCR activity
- Signal distribution source of prime time household viewing (i.e., cable, off-air, network)
- Top ranked syndicated programs
- Top ten network programs
- Top fifteen regularly scheduled network programs

The television program director has a number of additional aids that help determine program selection and placement. One is an analysis of available programs. For example, Leo Burnett, a major advertising agency, provides a yearly *Television Program Report* that lists pilots, shows for commercial buys, those in test-runs, and those in process, both on the network and syndication levels. Each entry discusses the program in terms of its characters, plot, type, and general orientation, and lists the production company, producers, directors and writers as well as the talent, where appropriate. Figure 4–1 shows the table of contents and some sample program descriptions of the *Television Program Report* for 1988–89.

Another aid is the product advertising analysis, showing the relationship of certain kinds of products and services to the demographics of viewers of certain types of programs. For example, the Cable Advertising Bureau publishes *Cable TV Facts,* a statistical breakdown of cable viewing habits, cable viewer demographics, viewing shares, products and services of special interest to cable viewers, and commercial placement organizations.

Target Audiences

While ratings and demographics are interdependent in developing program concepts and scheduling for both television and radio, there is a distinct difference between the target audiences of the two media. Radio, as it has developed since virtually all of its popular programming was preempted by tele-

TABLE OF CONTENTS

INTRODUCTION

This is the 31st annual edition of the Leo Burnett U.S.A. Television Program Report, prepared by the staff of the Television Program Department. It covers network television and syndication.

Information for this report was obtained from network program departments, as well as program producers and syndicators.

In the network section we include projects that have passed the development stage and are true pilots, commitments, and those that are scheduled as test-runs. We also include the mini-series to which networks are committed.

The projects are broken down by network and type. They are listed alphabetically and briefly described. In each write-up, the casting information has been included when available. Because it is necessary that we have a cut-off date for receiving information in order to have our report printed and delivered in advance of the networks' announcements of fall schedules, we sometimes do not have complete or possibly correct casting, titles are often changed and projects committed after the fact. We have each program cross-indexed by title, talent, and production company.

The report on the 1988-89 season lists one hundred and two developments in comedy and drama, five varieties and twenty-one mini-series.

In the syndication section the projects are broken down both by program type and basis of sale. They are listed alphabetically and are briefly described within the subheads. They have been cross-indexed by title and syndicator.

Note: As we prepared this report the television industry was involved in a writers' strike which could affect some of the pilots listed here. At press time some shows had already been put on the back burner and do not appear in this edition.

Bill Eckert
Vice President
Programming

COMEDY

TEST-RUNS

JUST THE TEN OF US

WARNER BROS. TELEVISION

Inspired by the utility teacher/coach character in the "Growing Pains" series, this spin-off centers on an average hard-working family man with a wife, five daughters, two sons and a baby on the way. When a budget cutback leaves him unemployed, he packs up his family and moves west to a new position in a private school for boys. Inarticulate, but occasionally poetic, he's a guy right out of the 50's doing his best to hold his own in the 80's. Coach Lubbock is played by Bill Kirchenbauer with Deborah Harmon as his wife, Elizabeth. Playing his daughters are Joann Willette, Brooke Theiss, Jamie Luner, Heidi Zeigler and Heather Langenkamp. Matt Shakman and (twins) Jeremy and Jason Krospjens are the two sons.

Director: John Tracy
Writers: Dan Guntzelman and Steve Marshall
Executive producers: Dan Guntzelman, Steve Marshall and Michael Sullivan

Figure 4–1 Table of contents and sample program descriptions of Leo Burnett's *Television Program Report.* (Courtesy of Leo Burnett.)

continued

ROSEANNE *CARSEY/WERNER PRODUCTIONS*

Commitment

The goal of this venture is to demonstrate that the role of being wife, mother and homemaker isn't always as easily and smoothly accomplished as is so often shown in television drama. There is another side of the picture in the real world where one finds the female equivalent of Ralph Kramden in existence. In this concept Roseanne Parker is one of them. She's a blue-collar mother frantically juggling kids, husband and home as well as holding down a tough factory job to get needed extra dollars. She proves that in the real world it takes a dedicated woman with a big dose of stamina to keep the family functioning as a family should. This concept is designed to showcase Comedienne Roseanne Barr in the role of Roseanne Parker. John Goodman portrays her husband, Dan. Her three daughters are played by Laurie Metcalf, Lisi Gorenson and Sara Gilbert with Sal Baron in the role of her son, D.J. Natalie West and George Clooney play her unsympathetic supervisors at the factory where Roseanne works.

Director: Ellen Falcon
Writer: Matt Williams
Executive producers: Tom Werner and Marcy Carsey

PILOTS

BABY ON BOARD
HART/THOMAS/BERLIN PRODUCTIONS

George and Sally are a career-oriented married couple who have been happy and content in their marriage without paying much attention to the fact that the years have been slipping by. Suddenly, the truth dawns on them. They always wanted a family and now they realize that they'd better get down to business before it's too late. That's what happens in the back story of this projected series in which Sally has her first child, Abigail, when she and George are in their early forties. They both quickly learn that their lives are changed forever. Sally is open-hearted, willing to make whatever adjustments necessary in order to properly accommodate the baby. But what is easy for Sally is difficult for George, much as he wants to oblige. Above and beyond the baby, new pressures enter their lives. They suddenly find themselves uncomfortable with couples who have new babies and are twenty years younger as well as being ill-at-ease with their older, childless friends. They also face the problem of Sally's mother and George's father who both worship the baby but can not stand each other. Since the grandparents can not help care for Abigail, George and Sally decide to hire a nanny. What a nanny they come up with! Lauri is a beautiful college student who is short on experience but long on common sense, thankfully. In any event, as the series would unfold, George and Sally step into a new life where a brand new baby in the house gives them more pleasure than they ever thought possible. This is a New York project to be shot entirely in New York. The group of actors in these roles are Lawrence Pressman as George, Jane Galloway as Sally, Joan Copeland as Marion, Larry Haines as George, Senior and Teri Hatcher as Lauri.

Director: David Steinberg
Writer: Peter Stone
Producer: Cathy Cambria
Executive producers: Carole Hart, Marlo Thomas and Kathie Berlin

MARY TYLER MOORE *(Working Title)* *MTM ENTERPRISES*

Commitment

Annie Block is an adventurous divorced woman who's deeply in love with her new husband, Nick McGuire, a handsome widower whose attitudes are often at odds with her own. Annie's an open-minded liberal eager to try exotic cuisines or experiment with pop psychology when disciplining her 12-year-old son, Lewis. Nick's a meat-and-potatoes man, an engineer who's more of a traditionalist at heart. Until she met Nick, Annie's life had revolved around Lewis and her job, working as the director of community relations for a New York borough president. Nick also has his hands full as the father of Debbie, a 9-year-old daughter, and Lenny, a 14-year-old son. To complicate matters, Red, Nick's conservative blue-collar dad, and Emma, Annie's liberal activist mother, can barely speak to each other without getting into a disagreement. Despite all the potential for domestic problems—the kids, the parents, and especially their own divergent attitudes—for better or for worse, Annie and Nick are ready to make a go of marital bliss. Mary Tyler Moore returns to television in the role of Annie with Edward Moore as her husband Nick. Eileen Heckart and John Randolph play Emma and Red and the children are Bradley Warren as Lewis, Cynthia Marie King as Debbie and Adrien Brody as Lenny.

Director: Beth Hillshafer
Writers: Elliot Shoenman and Paul Wolff
Producer: Paul Wolff
Executive producer: Elliot Shoenman

Figure 4–1 *continued*

DRAMA

TEST-RUNS

CHINA BEACH *WARNER BROS. TELEVISION*

This six-episode series focuses on the lives of a group of women who serve in Vietnam during the conflict there. It looks at the horrors of war through their eyes. Colleen McMurphy is a nurse scheduled to be returned to the U.S. after a long stint but realizes that her services in Vietnam are more valuable than any place else. She continues on at the hospital. Laurette Barber is a fun-loving U.S.O. performer and she, too, decides to stay after experiencing the realities of war up close. Cherry White, a kind and innocent Red Cross worker, came in search of her missing brother. She finds herself in a completely foreign and shocking situation but is determined to do her duty for her country. K.C. is short for Kansas City. She is also doing her duty in Vietnam. She is a beautiful, shrewd prostitute who realizes that this war-torn country is a potential gold mine. These dedicated women, working in an insane place and time, must be strong as they help the wounded and dying face the horror and pain of war. Other characters include Boonie, a friend of McMurphy's, still in the service but out of active duty and working as a lifeguard on the beach; Becket, a private who works in the morgue; and Natch, the helicopter pilot who files in the wounded and Colleen's love interest. Dana Delaney, Chloe Webb, Nan Woods and Marg Helgenberger are Colleen, Laurette, Cherry and K.C. Brian Wimmer, Michael Boatman and Tim Ryan are Becket, Boonie and Natch.

Director: Tod Holcomb
Writer: John Sacret Young
Executive producer: John Sacret Young

PILOTS

DESERT RATS *UNIVERSAL*
(Working Title)

Josh Bodeen was a rowdy youngster when he was growing up in Liberty County, Arizona. Now, at the age 26, he finds himself sheriff there, giving law enforcement a new definition in today's Southwest. He is assisted by the ever-cool Deputy Bones, who in earlier times was more often than not dragging Josh out of some brawl. When the standard-issue squad car burns out, Josh enlists a mechanical wizard named Owen to provide them with a torqued-out version and a souped-up helicopter. Swept up in all this is timid officer Mort Ledonka who tries to keep up with this fast-paced crew. With his unorthodox ways and his unique band of law enforcement officers who use an assortment of unusual techniques, Josh and crew fight crime of all kinds in this action adventure potential. Starring in these roles are Scott Plank as Josh, Scott Paulin as Bones, Mark Thomas Miller as Owen and Dietrich Bader as Mort.

Director: Tony Wharmby
Writer: David Chisholm
Supervising producer: Ken Topolski
Executive producers: David Chisholm and Bernie Kowalski

THE INCREDIBLE HULK RETURNS *NEW WORLD TELEVISION*

Bill Bixby and Lou Ferrigno reprise their roles as David Banner and his innerbeast, the Hulk. In this two-hour movie, David faces the opportunity to finally rid Hulk from his life. Fellow scientist Donald Blake seeks out David for help with his own "beast", the rambunctious thunder god, Thor. Thor is the spoiled son of the mythical Norse god Odin, who was banished to Earth by his father and must serve time doing good and helping mankind. In this backdoor pilot, while on an expedition along the Fjords, Blake stumbles into a cage and finds the hammer of Thor, which is Thor's strength. When Donald strikes the hammer, he awakens Thor who then becomes attached to Donald, although they remain individual beings. If an ensuing series results, it will focus on the continuing adventures of Thor who, along with Donald Blake, will carry out the sentence imposed by his father by fighting crime and injustice. Eric Kramer plays Thor with Steve Levitt as Donald Blake.

Director: Nick Correa (Pilot)
Writer: Nick Correa (Pilot)
Producer: Dan McPhee (Pilot)
Executive producer: Nick Correa and Bill Bixby (Pilot)

VARIETY

THE JOE PISCOPO SHOW *UNIVERSAL*

Popular comedian Joe Piscopo stars in this variety format featuring the familiar characters and impressions which he has developed on "Saturday Night Live" and his many beer commercials. Comedy sketches, musical guests and a supporting cast of regulars will round out the program.

Writers: Dennis Rinsler, Marc Warren, Carl Gottlieb, Neal Israel, Mark Ganzel and Don Rhymer
Producers: Marc Warren and Dennis Rinsler
Executive producers: Neal Israel and Carl Gottlieb

continued

THE JIM HENSON HOUR

Commitment — *HENSON ASSOCIATES*

Scheduled to begin airing in January of 1989 and hosted by Jim Henson, this series will feature a number of elements including the "Storyteller" fables, Muffet parodies of popular feature films and various new features such as "Infobits", which will humorously explain commonly used but little-understood topics such as the ozone layer. A special episode will celebrate the 20th anniversary of "Sesame Street." The series will be produced in New York, London, Toronto and Los Angeles.

Executive producer: Jim Henson

MINI-SERIES

THE FATAL SHORE

De LAURENTIIS ENTERTAINMENT GROUP

An epic tale, this six-hour series will trace the birth of Australia out of England's infamous convict transportation system, and the continent's subsequent transformation from an enormous jail for the English undesirables into a flourishing nation.

Telewriter: Ernest Kinoy

THE KENNEDYS OF MASSACHUSETTS

EDGAR J. SCHERICK ASSOCIATES

The whole and engrossing history of two great American families, the Kennedys and the Fitzgeralds, that converge powerfully on the American scene. This is a five-hour saga unfolding on a large canvas, one that befits two families whose eventual triumphs, strengths, weaknesses and mortalities could be inscribed in the chronicles of the world.

Telewriter: Diana Hammond
Producer: Sue Pollock
Executive producer: Edgar J. Scherick

SELMA, LORD, SELMA

CARSEY/WERNER PRODUCTIONS

A four-hour true story, this drama recaptures the explosive events surrounding the Civil Rights movement in Alabama as seen through the eyes of two young black girls. When their innocent cry for freedom rallies all of Selma to act, the girls are launched into an unexpected rite of passage, accompanied on their journey by Dr. Martin Luther King, Jr.

Director: Peter Werner
Telewriter: Alice Arlen
Producer: Caryn Mandabach
Executive producers: Tom Werner, Marcy Carsey and Henry Hampton

SYNDICATION

Two types of programs have been addressed in this syndication section: New programming created specifically for the syndication arena and off-network programming beginning its first year of syndication.

Program types include: comedy, variety, talk, other, specials, children's shows and drama. Movie packages, short inserts and drop-ins are not included. All listings are categorized as definite "go's" by the respondents.

KEY TO ABBREVIATIONS

W Weekly
O One-time-only mini-series
S Strip

C Cash
B Barter
\+ Cash +

NEW PROGRAMMING

CHILDREN

AMERICA'S FIRST TEAM (Animated) *ORBIS*

:30 (S) (B)
Famous athletes, as cartoon characters, helping children.

BOZO'S 3-RING SCHOOL HOUSE

LARRY HARMON PICTURES

:30 (S) (+)
"Themed Productions" featuring special entertainment, education game show segments with a live audience.

CHIP & DALE'S RESCUE RANGERS

BUENA VISTA

:30 (S) (B)
Walt Disney's Chip & Dale star in their own show.

Figure 4–1 *continued*

COMEDY

CANDID CAMERA *BLAIR ENTERTAINMENT*
:30 (W) (B)
Veteran Alan Funt is back in this all new show.

COLORS OF SUCCESS *M.K. THOMAS & CO.*
1:00 (W) (B)
A rhythm and blues club is the backdrop to this weekly narrative that incorporate celebrity quest into the storyline.

THE COSBY SHOW *VIACOM*
:30 (S) (+)
First year off-network.

IMPROV TONIGHT *PEREGRINE FILM*
:30 (S) (B)
Bud Friedman's Improv Tonight comedy club features new stand-up comics.

KATE & ALLIE *MCA-TV*
:30 (S) (C)
First year off-network.

DRAMA

AIRWOLF *MCA-TV*
:60 (S) (C)
First year off-cable (USA).

CAGNEY & LACEY *ORION*
:60 (S) (C)
First year off-network.

THE CAMPBELLS *ITF ENTERPRISES*
:30 (W) (C)
Scottish father and children live in the wilderness.

CRAZY LIKE A FOX *LBS*
:60 (W) (C)
First year off-cable (CBN).

GAME

GRANDSTAND
MAJOR LEAGUE BASEBALL PRODUCTIONS
:30 (W) (B)
Three teams compete for points by answering sports trivia as conveyed through baseball highlight clips.

LOVE COURT *ORBIS*
:30 (S) (+)
A cross between the "Newlyweds" and "People's Court", Pearl Bailey casts as the Love Judge.

THE NEW LIARS CLUB *FOUR STAR*
:30 (S) (+)
Same as original except for the new host Eric Boardman.

THE NEW FAMILY FEUD *LBS*
:30 (S) (+)
Host Ray Combs brings the original format back.

TALK

DON KING'S ONLY IN AMERICA
ACCESS SYNDICATION
:60 (W) (B)
Talk show hosted by boxing promoter Don King.

GOOD COMPANY *GROUP W*
:60 (S) (+)
Magazine format locally produced by KSTP, Minneapolis.

THE LARRY KING SHOW *TURNER*
:30 (S) (B)
Host Larry King converses with famous personalities.

VARIETY/OTHER

ALPHY'S HOLLYWOOD POWER PARTY
ACCESS SYNDICATION
:60 (W) (B)
Spotlight music and dance aimed at pre-teens.

BODY BY JAKE *FRIES ENTERTAINMENT*
:30 (S) (B)
Health and fitness with Jake Steinfeld.

DICK CLARK'S GOLDEN GREATS
TELETRIB
:30 (W) (B)
Five performances from Dick Clark's private collection of musical clips.

FOLLOW UP WITH EDWIN NEWMAN
SYNDICAST
:60 (0) (+) (2 Parts)
Documentaries dealing with human relations and social problems.

vision in the early 1950s, is local and fragmented. With the exception of full-service radio stations that have been able to attract a general audience, individual radio stations program to select audiences, targeting their formats to a distinct and, hopefully, adequately substantial and loyal segment of the population. Television, on the other hand, continues to seek as large a part of the general audience as possible. With the exception of a still relatively small number of specialty stations, such as those with predominantly or exclusively music video or religious formats, TV stations in large as well as small markets compete head to head for the same audience. Some smaller stations, mostly UHF, that cannot begin to compete with network affiliates in their market have experimented with specialized formats such as *sitcoms* (situation comedies) or game shows filling much of the schedule. Specialization has grown more rapidly in cable, where the proliferation of channels has motivated toward narrow radio-like programming, such as the Learning Channel, the Discovery Channel, all-sports channels, all-news channels, weather channels, and home shopping channels.

Perhaps the most difficult determination to be made by even the most accurate research survey is who is principally responsible for choosing the programs to be watched by a given household. The ratings attempt to measure who is watching; they do not measure who decided what is to be watched. One survey indicated that as much as 40% of the audience was watching only because someone else turned on that particular channel or program.

Interpreting the Ratings

In his book *The Broadcast Industry*, Robert H. Stanley suggests several key questions when interpreting audience research[1]:

- What does your local audience really want to see?
- Can you estimate in advance how much time you can devote without losing too much money to children's programs, women's programs, local sports shows, news documentaries and religious programs?
- After you've decided on categories, how do you select the specific shows in each format that will result in the maximum commercial sales?
- If you include programming for minority audiences, how do you determine whether the intended message is reaching the targeted viewers and eliciting some kind of positive reaction?
- To what degree do you make moral or political judgments when selecting programs, serving as an arbiter or even censor in the "public interest"?

Further interpretation of the ratings, says Stanley, goes beyond numbers[2]:

- Audience popularity is not necessarily identical with the top ratings. "Faithfulness of viewers, mail response, clamor for participation on the show, ability of air personality to identify with sponsor, personal appearances, etc." transcend ratings per se.
- Ratings dominance "should be closely tied to the above, but not necessarily in terms of frequency and reach." Positive interpretations facilitate the sales of national reps as well as local salespersons.
- Profitable return on a program is not necessarily tied to the ratings; a program's top rating may be offset by it costing too much and a station may turn over a larger profit with a second place program that costs less.

It is sometimes necessary to lose money on a particular program or daypart in order to maintain station image and goodwill that pays off at other times in the schedule. For example, if a station is in the Bible Belt, it may lose money on religious programming at certain times of the day, but unless it airs those programs it may turn away a potential audience that makes the difference between the financial success or failure of other programs.

Prime Time Access Rule

The prime time access rule (PTAR), noted earlier in this book, provided opportunity for independent producers and local programming to a greater extent than in the previous twenty years. But performance is not necessarily consistent with opportunity.

> It is the challenge of creating programs that will satisfy my management, my own creative tastes, and answer the needs of the community that draw me on. The opportunity created by the prime time access rule gave me a regular slot to fill with new local products. I know we tried, but we couldn't ever hope to fill seven hours weekly with original local programs.[3]

The challenge and opportunity for increased local revenue under PTAR has ironically prompted many stations to retain voluntarily some aspects of a regulation that the industry has successfully lobbied the FCC to rescind, ascertainment of community needs, as discussed in Chapter 1. Although no longer required to, many stations at the end of the 1980s were still surveying community leaders to determine the most important issues in the community in order to get ideas for local programming that would attract a respectable audience. Most stations, however, have settled into syndicated programs for the 7:00-7:30 P.M. or 7:30-8:00 P.M. period, usually stripping these programs Monday through Friday.

SCHEDULING

Programming is competitive. If a station is a network affiliate, the network does the programming during prime time, from 8:00–11:00 P.M., as well as offering programming the rest of the day. A station has the option of not clearing a given network show and substituting a syndicated or local program. It is not often that a station can find, produce, or afford a nonnetwork show that will garner as large an audience as the network offering, especially during prime time. During the rest of the dayparts (the broadcast schedule is divided into *dayparts* from 1:00 to 10:00 A.M., frequently called *other;* 10:00 A.M. to 4:30 P.M., Monday–Friday, *day;* 4:30 to 7:00 P.M., *early fringe;* 7:00 to 8:00 P.M., *access;* 8:00 to 11:00 P.M., *prime time;* 11:00 to 1:00 A.M., *late fringe)* a station does its own competitive scheduling.

A station's aim is not only to attract but to hold audiences through a broadcast day or at least through certain time periods. This continuation is called *audience flow.* Some studies have shown that as much as 80–90% of an audience has held through one show that continued for an hour, but only 60–70% of that same audience stayed on for the full hour if a different although similar type of program came on at the half-hour mark, and only half of the audience stayed if the new program was of a different type. The aim of the programmer is to come as close to 100% retention as possible.

One of the key methods for doing that is *counterprogramming.* What do the other stations present at given times? How are your programs holding up against theirs? Is their local interview program pulling your audience away? Would a sports program placed opposite their interview show not only win back your defectors but cause some of their regulars to defect to you? Is their evening news program beating yours? Should you schedule yours a half-hour earlier (or later) and counter theirs with a syndicated game show?

Lead-in programming (used as a base for the audience flow example above) is designed to attract a substantial audience for a given daypart with an especially appealing program, one that may even cost more than it brings in in commercial revenue (comparable to a loss-leader in retail merchandising). It is up to the program manager to capitalize on the heavy viewing for the lead-in with appropriate follow-up shows.

If a station has a high-percent audience flow during a certain daypart, that audience flow can be maintained with *block programming*. Block programming emphasizes the format genre that seems to be responsible for the good audience flow by stressing that format for the entire time period, such as several straight hours of sitcoms, or several game shows in a relatively short time period.

When a particular show establishes a loyal audience of sufficient numbers, it is stripped. *Stripping* is to air the same program (a specific sitcom, drama series, or game show) in the same time slot Monday through Friday. It establishes a habit and makes it much easier for people to tune in "Wheel of Fortune" every day at 4:00 P.M., "M*A*S*H" at 5:30 P.M., or "Hill Street Blues" at 11:30 P.M. Of course, stripping is possible only when there are a sufficient number of episodes of a given show in syndication, either reruns or in continuing production. One advantage of stripping is that there usually is a discount when buying large numbers of episodes of programs in syndication.

And if nothing else works, there's always the special to cut into a given program of the opposition. A sports special, a personality interview special, or a top rock star variety special, effectively promoted, can draw a large one-time audience to a station. It will then be necessary to use some of the other techniques to keep that audience.

Counterprogramming

When up against a top show, trying to compete head-to-head can be very risky. The odds are that a station will lose out by going against an established show of a popular star with the same kind of talent. Remember the failure of Joan Rivers versus Johnny Carson–David Letterman? Find an alternative with some basic appeals elements. A station can afford to be less competitive if a show is codeveloped with a syndicator or is produced locally and has syndication potentials. For a network, that is a large consideration. If a station wants a different kind of talent attraction, the station should get a proven talent that is not likely to lose whatever audience the station has by bombing right away. If it is a hosted show, the key person should be appropriate to the content of the show. The content should fit the time period. For example, FCC indecency regulations made some programming permissible after 12:00 midnight that was not permissible before that time. In order to provide an audience with an alternative to the ratings leader at the time period, a station should make sure that the audience knows what comes when. On a late night show, for instance, segments would be set up in such a way that the audience could tune in materials that do not duplicate what they might otherwise be watching on "The Tonight Show" or "David Letterman" at a specified time.

Squire Rushnell, vice president of late night and children's television at ABC Entertainment, offers the following guidelines for late night shows: the talent must be attractive; the talent must be appropriate to the show; the content must be appropriate to the time period; the show must be different from anything else on the air at that time. Even when the format is the same, Rushnell states, the difference in personalities can make different shows, as exemplified by Oprah Winfrey and Phil Donohue.

Daytime shows are difficult to come by. ABC did an audience survey a few years back and found that the largest and most consistent interest, aside from the standard soaps and game shows, was in how-to programs, in depth, regarding such activities as home repairs. On that basis ABC developed the "Home" show. As of this writing its acceptance or rejection has not yet been determined, although ABC considered initial critical response in 1988 to be "phenomenal." ABC attributes this response to the show not being a lecture or a talk show on the subject, but "dramatizations of real problems with real people."

Soaps (soap operas or daytime dramas) require empathy of the audience with the characters, and vice versa. What audience needs are there to identify with this year, this month, this week? Game shows are essentially the classic types that may come and go, but are always there in one form or another. Prime time game shows attract the same audience as their morning counterparts. There is always a need for a new, innovative game show. The old ideas seem to be used up. At the beginning of 1988, for example, the last new network game show to go on the air was "Family Feud," which premiered in 1976.

Demographics are critical when counterprogramming—not only the demographics of the service as a whole, but those of the viewers of the competition's programming. For example, suppose a station's prime time access program consistently has been finishing in third place in a city with three network affiliates and several independent UHF stations. It will be necessary to counterprogram with something new and/or different that not only will draw nonviewers, but will draw off at least some of the viewers of the two leading programs. The demographics of each of the other program's viewers is the key. Suppose opposition program No. 1 is found to have very strong support from women twenty-nine to forty-five years of age, but is weak with twenty-five to thirty-nine-year-old males of an upscale educational and economic status, and weak with youngsters fifteen to eighteen years old. Suppose opposition program No. 2 is found to have its strongest support from both women and men aged eighteen to twenty-nine and from youngsters twelve to fifteen years old. Obviously, programming geared to upscale males twenty-five to thirty-nine years old, youths fifteen to twenty-five years old, and females twenty-five to twenty-nine years old has the best chance of cutting into program No. 1. Content geared to women twenty-nine to forty-five years old and youths fifteen to eighteen years old could impact on program No. 2. Viewers over forty-five years of age do not constitute as active a purchasing group as the younger tiers. One might decide on trying to push past No. 2 and take second place. One might decide to go for it all and come up with a program that will appeal to upscale males twenty-five to thirty-nine years old and to young teenagers. On the network level this would suggest several possibilities, including appropriate sit-com plot and character shows and, fairly obviously, a sports show. On the local level this might be limited to sports, or to male-interest public affairs programming, or even to audience participation or game shows that concentrate on the demographic groups being wooed.

The Station Representative

The role of the station representative is discussed principally in terms of advertising in Chapter 3. By virtue of their knowledge of many markets, the stations in a given market, and the past and current success of various programs, many station reps serve as unofficial program consultants to their client stations. Higher program costs and increased competition for audiences because of the growth of cable require careful program selection by program directors. A national rep can advise the client on the success of a show or series (including comparisons with similar markets), on the program's ratings in other markets and against competing programs in the client's market, on the cost versus potential income of that program, and on its record with specific demographic groups. The rep frequently can find a program based solely on the client's demographic potentials in the given market. The rep has access to national, regional, and local demographic and rating data frequently not available to an individual station. In addition, the rep's contacts with program sources, including networks and syndicators, provide information about given programs not usually obtainable by a station.

In some instances the station rep is a de facto program director, setting up acquisitions and even program schedules, and determining counterprogramming, placement, and frequency. Although affiliates rely on network offerings for their prime time programming, they still must make individual station decisions on access and nonprime time. Independent stations sometimes rely on the station rep for advice on their entire program schedule—with the exception, of course, of locally originated programs. In large markets (i.e., large stations), the station rep deals directly with the program director. In medium and small markets, the station rep frequently deals directly with the station manager; the program director of that station simply implements the program decisions determined by the station manager and rep.

In the 1980s, the role of the station rep grew to encompass more advisory authority in the area of news as well as entertainment programming, and to include management structure and production values as well as the format of the news show. The station that hires a major station representative firm may find that it has employed not only a sales consultant but a programmer as well. The programming advice is, of course, designed to raise the ratings of the given station. This in turn makes it easier for the rep to sell that station, and at higher rates, resulting in higher commissions for the rep and the firm.

SOURCES OF PROGRAMS

Stations have three major sources for their programs: networks, their own local productions, and syndicators. Formal affiliation agreements with networks provide access to their programming. Locally, a station can produce a given program in-house or arrange for an independent production company to make it for the station. Syndicators are independent producers and/or distributors, ranging from small organizations that have only one or a few shows available, to large Hollywood studios that provide a number of prime time series and feature-length movies. Cable systems use the same sources.

Networks

Stations affiliated with networks get about two-thirds of their programs from the networks. Independent stations, of course, must seek other sources. In a community with a small number of stations, a station may enter into an agreement with more than one network, as long as that network is not being carried by another station in the same service area. This provides the public with programs from all three major networks even if there are fewer than three stations that can be received. Stations have the option of clearing (or refusing) any given program. Even the most popular program nationally may be deemed unsuitable or lacking in appeal in a given community, and the affiliate in that community may simply refuse to carry the program. Sometimes pressure groups, as discussed earlier, convince a station to refuse to clear one or more episodes of a show. In some viewing areas, with large Catholic populations for example, stations have been convinced to drop discussion programs and even sitcoms that deal with abortion.

The network programs that the affiliates carry include the commercials placed with the networks by national advertisers. In return for such carriage, the affiliate receives a fee from the network. The affiliate also has the option of inserting local commercials (which, in fact, return more revenue to the individual station) at certain breaks in the program and in transitions between programs. Cable systems, incidentally, operate on a different basis. They must pay a fee to the cable production service or cable network in order to carry its programs.

Public broadcasters operate much the same way. There is only one "real time" public television network comparable to CBS, NBC, and ABC: the Public Broadcasting Service (PBS). Through a series of screening and bidding procedures, the PBS affiliates agree to pay a specified fee for a given program or program series (essentially, the total cost of the program divided by the number of affiliates carrying it, which creates a different fee for almost every program, no matter how similar in nature and length).

Commercial networks produce some of their own programs and contract out for production to other sources, specifically film studios and independent producers.

Many network programs that are placed against the highest rated show for a time period fail, not because the networks do not back them, but because the local station decides that the programs will not offer reasonable competition to the ratings leader and does not clear them. For example, ABC's "Our World," which was counterprogrammed to "The Cosby Show," was preempted by many affiliates in order to put on a totally alternative program, such as a game show, that might lure some of the "soft" Cosby viewers. Lack of sufficient clearances inevitably results in the switching of a show to another time or in its cancellation, no matter how good its quality.

Film Studios

After fighting television in its early years, film studios decided to join it because they couldn't beat it. With the development of videotaping, live television was no longer necessary, and drama production moved from New York to Hollywood, where the film industry had more experience in the new TV techniques of editing. Film studios, in fact, now get more of their income from producing for television (and this includes cable and home videos) than from films produced for movie theatres.

Independent Producers

After many years of being shut out from prime time television, independents made a breakthrough with the success of innovative programs such as those from Norman Lear's independent production house. (Some effects of PTAR are discussed elsewhere in this chapter and in Chapter 1.) By the late 1980s, network programming was about equally divided between that coming from big Hollywood studios and that from independent producers.

Syndication

The business of syndication has become so integral to television that the FCC found it necessary to develop rules governing it. According to the FCC, syndication is "any program sold, licensed, distributed, or offered to television stations in more than one market within the United States for non-interconnected television broadcast exhibition, but not including live presentations . . . " Distribution companies get the rights to a series from the producers and then contract with individual stations. Most of the syndicated programs are reruns, with many independent stations finding that a rerun of a popular show can be an effective counterprogram to a mediocre network program in that same time slot. An increasing number of syndicated shows are *first-run;* that is, they have not been and are not intended for network carriage, but go directly to independent stations. There is nothing to prevent network affiliates from using first-run syndicated programs to beef up a given time period when its network offers something that is weak in its service area. First-run syndication includes sitcoms, game shows, entertainment specials, public affairs programs and, in fact, all of the formats one sees in prime time as well as in other dayparts.

Not only did PTAR open the door for increased syndication, but the quality and appeal of some syndicated shows are so great that they frequently preempt prime time network shows at individual stations. The rights to syndicated shows are purchased by the station, which gives that station the available commercial time slots to sell on its own. If it fills all or most of the commercial time, the station does better financially than with the fee income from a network program. Stations with limited program budgets and a relatively weak or even moderate advertising market can arrange for barter syndication. One arrangement is for the syndicator to fill much of the commercial time with national ads that it has sold and to give the program free to the station in return for broadcast time. The station makes its money by selling the remaining commercial time on the show. Variations of barter include some fee payment by the station for programs in which substantial amounts of commercial time have not been sold, and the designation of advertising spots in later and different programs on the station in exchange for the show.

Local Production

Virtually all local production is in the area of news and public affairs. Network affiliates usually have local news shows, some several times a day: morning and/or noon and/or early evening and/or night. Independent stations that have the resources to distinguish themselves from other independents in the market frequently do a local news show that they place in a noncompetitive slot with network affiliates (i.e., 10:00 P.M. versus an affiliate's 11:00 P.M. news show that has network feeds).

The high costs for many syndicated shows, including full-length movies, limit the budgets of independents and affiliates alike for local production. About one-fourth of a station's total expenses, on the average, goes for programming; not much is available to produce one's own shows. The competitive innovative station that has the monetary resources may attempt a local prestige show, usually during the access period. Sometimes such a program is competitive enough to draw advertisers and pay for itself, and even make money. Even when it is not profitable it can be of value by heightening the station's image, sometimes by winning local Emmy awards for local production. A principal example of this kind of program is "Chronicle," on WCVB, Boston, a magazine-type news program aired every night during the access period.

COSTS

Many stations would like to produce their own programs, programs that are more ambitious than a standard newscast or an occasional feature. The problem is cost. While a syndicated program may not be altogether satisfactory because of its lack of special appeal to the demographics of a station's potential viewers, its relatively lower cost compared to a station-produced show offsets the disadvantage. In most cases even a syndicated show that may be more expensive than a locally produced program (a game show, for example) will draw a greater audience because of its appeal to the lowest common denominator.

But the local station does not have to worry as much as do the networks, with their large and sometimes fixed programming staffs. When revenues begin to flatten out, as they did in the mid-1980s when cable began to draw off broadcast viewers and make initial inroads into available advertising dollars, the costs of production do not follow the same trend. Production costs follow the economy and go up with it, even if at times only slightly. In the mid-1980s, for example, an episode of "Miami Vice" cost about $1.5 million to make. Only about two-thirds of that was in direct costs. Producers tend to blame unions for the additional costs, citing the approximately 35% benefits on top of salaries specified in union contracts. Conversely, the unions cite the production studios' usual addition of about 30% to the actual production costs for allegedly amorphous overhead. In trying to cut the costs of programs, networks look at producers and unions and usually disregard their own overstaffing and expense costs.

Production costs vary according to the complexity of the program, the degree of voluntary participation (i.e., college television majors), the cost of living in that geographical area, whether the station is union or nonunion, and similar factors.

Budgets are developed for line-by-line items and divided into *above-line* and *below-line* categories. Above-line costs in production include writers, directors, producers, performers, and other talent and other nontechnical personnel. Below-line costs include facilities, equipment, camera operators, costumes, scenery, lighting, sound, props, editing, processing of film or tape, and other technical items. On the average, 55% of a program's budget goes to below-line items, and 45% for above-line.

The actual costs of the technical aspects of a typical week's (5-days, 55-hours) production do not seem so formidable as noted in Figure 4–2, in 1986 figures. But when you consider that these figures do not include the production personnel (director, cinematographer, production manager, and art director) nor the benefits or fees noted earlier, you can see how an hour show can cost well over a million dollars.

People in the industry have begun to realize that the day of the "blank check" is over. Increasing costs, cable competition, and tougher economic times in the late 1980s spurred a move toward "affordable creativity," according to N. W. Ayer vice president Robert Igiel. Production costs are going to have to come under control.

The need to sell a substantial number of spots at substantial sums in a given time period becomes obvious. In October, 1987, the average cost of a 30-second prime time spot on the three major networks was $121,860. Each network can accommodate approxi-

Figure 4–2 Sample weekly staff costs for a TV production. (Courtesy of *Forbes.*)

Job category	*Union rate*	*Personnel*	*Nonunion rate*	*Personnel*
Assistant directors	$5,617	2	$3,186	2
Camera crew	8,694	4	4,897	3
Clerical & craft service	5,915	5	4,661	5
Costumes	5,520	3	4,542	3
Day players	10,240	10	2,360	10
Drivers	30,340	22	19,824	16
Electrical & lighting	13,872	5	6,726	4
Extras	10,360	50	2,950	50
First aid	1,286	1	885	1
Grips	7,241	4	6,608	4
Hair	3,573	2	2,950	2
Location manager	1,770	1	1,770	1
Makeup	5,641	3	4,956	3
Props	5,517	3	4,130	3
Script supervisor	2,070	1	1,652	1
Sets	9,140	6	5,240	4
Sound	7,155	3	3,540	2
Special effects	5,520	2	4,720	2
Stunts & stand-ins	11,120	5	3,304	5
Total	$150,591	132	$88,901	121

mately 1,144 prime time 30-second commercials each month—a total potential revenue of over $400 million, which could translate into over $5 billion for the year. Effective control of costs, most specifically program costs, could result in huge profits. The sales of TV stations for increasingly higher millions and millions of dollars attests to the profitably of the industry. Naturally, the cost of a given spot varies greatly with its placement in terms of show and time. Figure 4–3 shows the variations during 1987–1988.

PROGRAM TYPES

News

News is the most important locally produced program. With the advent of miniaturized electronic news gathering (ENG) equipment, there are no longer technical bars to any station gathering the local news and editing it into a local news show.

In a small market, there is limited competition, except for cable imports from distant cities. If a station has all of the local news coverage to itself, it will probably choose the simplest format, one that is homey and down-to-earth to reflect the likely homogeneous quality of its viewers. In a small market a station is probably affiliated with one or more networks, and the placement of the local news program usually depends on the placement of the network news show. In addition, if the station is getting any network feeds during the news show, their nature will affect the format.

In a large market it is a different ball game. A station must compete with other local news shows. It will likely want to make its show similar in format to its network's news program, if it is an attractive one, to maintain audience flow. At the same time the station will have to come up with some distinctions between its news show and those of its competitors. The usual difference is in anchor persons. Quite often the show with inferior coverage but with more attractive anchors will outrate a better news show. It is interesting to note how frequently stations in a medium-to-large market change anchors. A station will also use its promotion department to promote the personalities even more than it promotes the news format or coverage.

In addition to personality differences, exciting formats can draw audiences. Some stations use their ENG equipment to special advantage (see Chapter 5) and invest in live field coverage. Other stations try to combine entertainment with the news, hoping that will draw viewers. For example, a station whose coverage area includes urban, suburban, and rural areas, and even crosses state as well as county lines, might originate the news from a different site over a period of nights, showing the audience something about that remote (far away as well as out-of-studio) area, involving the news personalities as quasi-tour guides.

Unfortunately, entertainment has permeated content as well as format. Whether it is an attempt to find any gimmick that can compete successfully or whether it is the result of increasingly ill-prepared journalists, American television news, with a few exceptions, such as the McNeil-Lehrer report on PBS, has moved more and more into entertainment since the days of Edward R. Murrow. The short time allotment for any given news story (90 seconds of "in depth" coverage), the emphasis on visual action, the concentration on upbeat stories, the downplay and even omission of downbeat critical stories, and the attempts at jovial chit-chat among anchors have virtually turned the news business into show biz.

Cable, especially the Cable News Network (CNN) and C-Span, has outstripped the broadcast networks and most local news stations with coverage of news in depth as well as in breadth. Naturally, no broadcast station can afford to spend a substantial part of its broadcast day on news alone, as does

Day	Network	7:00 (ET)	7:30	8:00	8:30	9:00	9:30	10:00	10:30
MONDAY	ABC			MacGyver $90.330		Monday Night Football $195.000			
	CBS			Kate & Allie $102.640	Frank's Place* $100,350	Newhart $149.190	Designing Women $124.790	Cagney & Lacey $115.440	
	NBC			ALF $149,500	Valerie's Family $161,000	NBC Monday Night at the Movies $150.000			
TUESDAY	ABC			Who's the Boss? $197,660	Growing Pains $228.350	Moonlighting $249.190		thirtysomething $125,460	
	CBS			Houston Knights $87,850		Jake & the Fatman $91,520		The Law & Harry McGraw $76,080	
	NBC			Matlock $103.000		J. J. Starbuck $82,000		C[illegible]e Story [illegible]8.000	
WEDNESDAY	ABC			Perfect Strangers $138,110	Head of the Class $159,290	Hooperman $146,780	The "Slap" Maxwell Story $127,090	Dynasty $113,540	
	CBS			The Oldest Rookie $75,360		Magnum, P.I. $140,790		The Equalizer $134,520	
	NBC			Highway to Heaven $109,000		A Year in the Life $134,500		St. Elsewhere $131,000	
THURSDAY	ABC			Sledge Hammer! $38,740	The Charmings $45,860	ABC Thursday Movie $79,830			
	CBS			Tour of Duty $67,590		Simon & Simon* $85,770		Knots Landing $127,050	
	NBC			The Cosby Show $369,500	A Different World $276,000	Cheers $307,000	Night Court $257,500	L.A. Law $205,000	
FRIDAY	ABC			Full House $79,780	I Married Dora $68,660	Mr. Belvedere* $89,960	Pursuit of Happiness* $79,090	20/20 $80,000	
	CBS			Beauty & the Beast $68,900		Dallas $139,470		Falcon Crest $99,700	
	NBC			Rags to Riches $91,500		Miami Vice $171,500		Private Eye $124,000	
SATURDAY	ABC			Ohara* $55,680		Sable* $37,070		Hotel $83,480	
	CBS			CBS Saturday Movie* $57,000-$100,000				West 57th $48,120	
	NBC			Facts of Life $107,000	227 $111,500	Golden Girls $182,500	Amen $135,000	Hunter $121,500	
SUNDAY	ABC	Disney Sunday Movie $78,734		Spenser: For Hire $64,530		The Dolly Parton Show $138,210		Buck James $94,370	
	CBS	60 Minutes $160,300		Murder, She Wrote $140,560		CBS Sunday Night Movie $121,020			
	NBC	Our House $92,500		Family Ties $219,500	My Two Dads $189,000	NBC Sunday Night at the Movies $155,000			

Figure 4–3 Varying spot costs for different network prime time shows. (Reprinted with permission from *Advertising Age,* January 4, 1988. Copyright 1988, Crain Communications, Inc.)

CNN, considering that the lowest common denominator is the ratings factor in our advertiser-supported television system.

Perhaps some network or station will one day be able to compete successfully or be willing to invest in serious public service by going in the other direction: hard news in depth with competent analysis and a willingness to cover all sides of controversial issues.

While affiliates get some of their news from their networks and most can afford to produce their own locally, independent stations are in a difficult funding position. As noted earlier, some manage to support a local news show. All of them, if they wish, can offer some news even if they cannot afford to produce their own. International and national news is available to independent stations from CNN and from the Independent News Network (INN). Individual news stories, that is, not a complete news telecast, are available from the Associated Press (AP) and United Press International (UPI), and can be used for short news shows such as five-minute news updates. In addition, a station can record feeds from its network source or from independent sources such as INN and UP-ITN (UPI and ITN—Independent Television News of England) and integrate the material into its own news program at a later hour.

Sports

ABC paid over $300 million for the rights to the 1988 Winter Olympic Games in Calgary, Canada. Not unexpectedly, viewer interest in watching the Olympics gave ABC the ratings advantage for that period. Even if a network loses money on some sporting events, the ratings increases they generate and the audience flow that may be siphoned into other presentations of that network and its affiliates are worthwhile compensations. Of course, the financial benefits from being able to insert local commercials at specified places are significant to the affiliate, considering that many sports events are watched by some 35–50% of television sets in use at any one time.

How is the success of network sports translated into local sports for both affiliates and independents? Local stations competitively bid for the rights to local teams, including major league franchises in large markets. Independent stations find they are frequently able to outbid their usually wealthier affiliates, not necessarily by offering more money, but by being able to clear more time for the presentations of the contests, especially in prime time, where the affiliates simply cannot preempt on a regular basis highly rated network entertainment shows.

Contracts for local sports teams differ, with each station and team working out their own special arrangements for compensation, the number of games to be covered, home versus away coverage, scheduling of timeouts and other breaks in the contest for commercials, promotion details, blackouts (including any requirement that a home game must be sold out in order for it to be carried on a local station), and the selection of the game announcers.

Local stations that lose out on professional sports or quasi-professional collegiate football and basketball coverage still have an opportunity to attract smaller, but select and frequently loyal, audiences by televising less demanded sports (i.e., swimming and soccer) and local high school major sports. Local rivalries frequently generate high ratings for a small, independent UHF station. Athletic events are extremely popular, and cable all-sports networks have become so successful that some of them even have become pay channels.

Public Affairs

Federal Communications Commission deregulation eliminated the requirement for a minimum amount of news and public affairs programming. Nevertheless, many stations continued public affairs programming. While few stations are going to attempt "60 Minutes"-type documentaries, which are extremely expensive to produce, some produce so-called minidocumentaries (minidocs) of some five to seven minutes each. WCVB, Boston, has produced many high-quality minidocs on its daily "Chronicle" show. Public affairs programs may be in the form of discussion panels, interview panels, talk shows, speeches, editorials, coverage of political events, local government meetings, and the like. In most cases these are not too expensive to produce and may entail only a few hours of field or remote or studio work, in addition to air time. Talent, outside of station staff, usually costs nothing, consisting principally of public figures or just plain citizens.

In this program area, too, cable has been able to outstrip broadcasting, with all-public-affairs channels available at all times. On the national scene, for example, C-Span provides continuing coverage of the Congress and of other federal agencies and operations. On the local level most cable systems provide a governmental access channel that carries ongoing coverage of the city council and other municipal departments.

Movies

Feature films are a staple of both network affiliates and independent stations. Networks counterprogram with prime time movies, some of them recent Hollywood hits, others made-for-TV with name performers. Earlier in the day the affiliates program films that they have obtained through syndication. Many smaller independent stations rely principally on films to fill their schedules. The largest programming expense for most stations is for films. In a large market the rights for one showing of a feature film can cost $50,000 and more; even in smaller markets the cost is substantial. When buying packages—30, 100, or 200 titles or more—the cost per film is less, and sometimes a package averages only a few hundred dollars per showing. Granted, these are not films likely to attract large audiences, but they complete schedules and can draw enough to make a profit even with only a minimal number of commercials. Compare these prices to the $15 million ABC paid in 1986 for the rights to show "Ghostbusters."

Although made-for-TV movies are still produced principally for networks, more and more are being sold to cable. Some made-for-TV films are in multiple segments for airing over a period of several or more nights; these are the *miniseries*. Local affiliate and independent stations not only are able to buy syndicated made-for-TV movies that already have been shown on broadcast or cable, but they can buy made-for-TV movies that did not make network or cable. These sometimes serve independents as good counterprogramming to the already known Hollywood films on network affiliate prime time.

Other

Virtually no local origination is done for other program types. Most other formats are either obtained through syndication or distributed to the community through alternative systems, such as public television and cable.

Sitcoms and Drama

Syndicated series span virtually the entire history of television, some available series going back 35 years and more (i.e., "I Love Lucy") and some as fresh as yesterday's prime time shows (i.e., in 1987 "Hill Street Blues" began syndication throughout the country as soon as it completed its network run). In some markets independent stations wage a counterprogramming battle of sitcoms. Producing a local drama is virtually out of the question because of costs, although a station may occasionally produce a special docudrama or carry a play produced by a local community or college dramatic group.

Game Shows

The local station cannot match the syndicated programs and, at best, may carry an occasional special or short series of a special nature (i.e., a citizen competition tied into a merchandising promotion, or a quiz contest between local high school or college teams).

Music and Variety

Although they are extremely popular in some markets, music videos are very expensive to do well and are carried through syndication or through music television (MTV) stations. A station occasionally may carry a community orchestra performance or a touring rock concert. Some stations produce late night variety shows using local talent, with production and technical crews (if a station is nonunion) from local college media departments.

Children's Programs

Although many network children's shows, such as "Kukla, Fran, and Ollie," started at local stations, the costs of production and the easy availability of network and syndicated "kidvid" shows make local production virtually a thing of the past. Depending on its public interest commitments and its budget, a station can develop a children's show oriented to the needs and interests of children in that viewing area. Some have been highly successful. For example, in 1988 Boston's CBS affiliate, WNEV-TV, counterprogrammed syndicated animation and local news programs on other stations 6:00 to 7:00 A.M. weekdays with a locally produced children's show, "Ready to Go," using a "Today" show type of format. The program included news, games, weather reports, adventure, health information, trivia, and other features oriented to six to twelve year olds. The station replaced an exercise program and a sitcom with the children's program.

Religious Programming

Many stations devote time to local church services and evangelists. Providing free time for one denomination, however, usually means having to provide free time to others, and when stations began to charge a fee to religious groups for program time, many of these groups began to move to the less expensive medium of cable. Many TV stations are owned and operated by religious organizations and devote their programming to their denominations. Many stations still provide free time, even though they may incur substantial losses by doing so. As one station manager said, "We're located in the Bible Belt and can't afford not to do so."

Educational

Formal educational programs are found principally on public television stations. Few commercial stations still carry college credit courses or the instructional materials used in elementary and secondary classrooms. Many commercial stations provide informal educational shows, such as the opening of a new exhibit in the community art museum, although technically such programs would fall under public affairs and not education. A number of features produced by public affairs departments, such as discussions of health services, environmental pollution, elderly housing, and similar topics, can be considered educational.

THE FUTURE

Television programmers not only have to program competitively against other stations, but increasingly against other distribution systems. Cable and VCRs are driving television stations onto roads where they have never been. Broadcast television viewing began to decline in the mid-1980s as cable penetration and home videos grew. Some critics believe that viewers are moving away from general programming and are seeking more targeted programs.

On one hand, cable has offered increasing competition to broadcast stations for the audience and for programs and program series. The latter raises the ante for the purchase of shows. On the other hand, the needs of cable for more programming widens the syndication market, not only for independent producers, but also for the networks and individual stations that have produced programs. "Miami Vice" is an example of both sides of the coin: syndication rights were sold to the USA cable network rather than, as likely would have been the case up to the time national cable penetration passed 50%, to broadcast stations.

One of the advantages of cable is that it is not bound by daily schedules, as is broadcasting. It is much easier for cable to spot a show wherever and whenever it seems best to run it. The proliferation of channels and the nature of cable programming at its present evolvement permit entire channels of a specialized nature and the repetition of popular programs at selected times during the week. The lucrative home video market permits a viewer to rent a virtually first-run feature film and offers a greater variety of choices than is usually available on broadcast or even cable TV.

By the 1990s broadcast television programmers are going to seriously have to evaluate and reorient their programming if broadcast stations are going to continue to grow and, perhaps, even survive.

There is a further bottom line.

Using appropriate techniques and with a decent budget, one can become a successful scheduler as measured by the ratings. But one must not let the techniques block out the larger responsibility: the public interest, convenience, or necessity. What a station programs, and for which audience, has an ethical as well as monetary base.

NOTES

1. Stanley, Robert H. *The Broadcast Industry*. New York: Hastings House, 1975, p. 98.
2. Ibid., p. 96.
3. Ibid., p. 97.

BIBLIOGRAPHY

Books on media research listed in the bibliography to Chapter 3 are applicable, as well, to programming decision-making. Many of the periodicals listed in the bibliography to Chapter 1 deal with program trends and evaluations of programming; *TV Guide, Variety,* and *Broadcasting* provide current updating.

Beville, Hugh Balcom. *Audience Ratings: Radio, Television and Cable*. Hillsdale, NJ: Lawrence Erlbaum, 1985.

Blum, Richard A. and Robert D. Lindheim. *Prime Time: Network Television Programming*. Boston: Focal Press, 1987.

Brown, Les. *Encyclopedia of Television*. New York: Zoetrope, 1982.

Clift, Charles III and Archie Greer. *Broadcasting Programming: The Current Perspective*. Washington, DC: University Press of America. Annual compilation of pertinent articles.

Eastman, Susan Tyler. *Broadcast/Cable Programming: Strategies and Practices*. 3d ed. Belmont, CA: Wadsworth, 1989.

Gitlin, Todd. *Inside Prime Time*. New York: Pantheon, 1985.

Howard, Herbert H. and Michael S. Kievman. *Radio and TV Programming*. New York: John Wiley, 1983.

5

Technical Aspects: Studio, Control, and Transmission

Elizabeth Czech-Beckerman

Television studio equipment becomes more and more complex and varied each decade, with automation becoming a dominant factor as the twenty-first century approaches. Nevertheless, the human being still remains the key. There is still need for an operator to understand the delicate nature of the equipment being handled. A competent operator knows not only which buttons and switches to push, but also understands the interdependence between the equipment and the total electronic process.

The beginning operator usually handles only one position, at the simplest level. In television, however, each person is part of a team and the individual who shows knowledge of the total process not only functions more efficiently, but is likely to be promoted sooner.

This chapter is an overview of the complete technical operation, from shooting the initial presentation (in the studio or field) through the transmission of the picture and sound to television receivers. It examines basic types of equipment found in control rooms, studios, remote vehicles, and auxiliary systems, and analyzes how television uses the electromagnetic spectrum for video and audio. It also describes the state of present technology and expectations for future technological change. One should keep in mind that although the term *broadcast television* is used here, similar equipment is used in corporate and industrial television, cable, closed circuit, satellite, and other forms of telecommunications.

STANDARDS

Equipment

The FCC, among its many responsibilities, sets technical standards for manufacturers to meet in order to produce type-approved equipment that ensures compatibility of systems and a resulting quality broadcast system. Since its emphasis on deregulation, beginning in the late 1970s, the FCC has lessened the rigidity of its equipment standards and takes the position that the marketplace should decide which new equipment is or is not compatible with existing systems.

In the U.S., the Society of Motion Picture and Television Engineers (SMPTE) also sets equipment standards, but the SMPTE cannot regulate, only recommend. For nations of western Europe the European Broadcasting Union (EBU) operates an elaborate program exchange system called *Eurovision* and has the authority to insist that its members adhere to approved standards for their broadcast equipment.

Therefore, equipment manufacturers have formal and informal guides for complying with standards that provide quality video and audio and compatibility with other systems.

Video

In 1939, when telecasts were experimental, the FCC withheld permission for full-scale commercial operations until the industry could agree on engineering standards. The National Television Systems Committee (NTSC), representing fifteen major electronics manufacturers, recommended standards for black and white television, including 525 lines per frame and 30 frames per second, which the FCC adopted in 1949 and which are still being used in the United States. The same NTSC standard is also in operation in Canada, Mexico, and Japan, which use 60 cycle current as we do in the United States. Brazil is the only other country with 60 Hz house current, but it uses phase alternating line (PAL)-M (a modification of the PAL system described below).

Sequential color with memory (SECAM) was developed in France and is used in about twenty countries, including France, Russia, and much of eastern Europe. Along with PAL, SECAM was developed for countries having 50 cycle house current, resulting in twenty-five frames per second and 625 horizontal lines, thus offering better resolution than NTSC or PAL-M. (*Resolution* means a better, more defined picture.)

Developed in Germany, PAL is used in thirty-six countries, including West Germany, England, and Holland. SECAM, PAL, and PAL-M differ from NTSC in the way the color signals are combined with the brightness signal. Fortunately, it is possible to convert a program from one standard to another by running a signal through a standards converter and recording the converted signal on a duplicate tape. It should be noted that in some countries a standard has been chosen more for political reasons than electronic; for example, when a nation does not want its signal to be picked up by its neighboring countries or does not want other countries' signals invading its territory.

Television networks in the U.S. require compatible equipment standards; the existence of multiple standards would create many logistical problems for the far-flung network news and sports operations. Field crews often need to use editing facilities of an affiliate to put together a complete story to be fed back to New York. Multiple standards would also create problems for broadcasters who are part of national and international communications projects, such as the Olympics or major news events abroad. Standards, therefore, are established for a variety of reasons and have an important impact on the development of equipment and technologies.

SPECTRUM

In order to understand how television equipment works and how its signal is distributed, one must first understand some things about the electomagnetic spectrum. Using electricity and electrical currents, television works in much the same way as does radio. Television's and radio's video and audio signals are changed into electomagnetic (invisible light) waves that are sent through the air. Spectrum order must be maintained so that every station does not interfere with every other station, as they did in the early days of radio.

Because frequencies do not respect national frontiers, an international authority known as the International Telecommunications union (ITU) was in operation as early as 1865. Today ITU functions as a specialized agency of the United Nations and holds World Administrative Radio Conferences

(WARC) among all governments, one purpose being to minimize the adverse consequences of overlap and interference among broadcasting signals in different countries in order to avoid chaos on the spectrum. ITU achieves this by allocating bands of frequencies to each service (i.e., television, radio, short wave, marine navigation, in addition to noncommercial areas such as aeronautical, maritime, amateur "ham," international, citizens band, land-mobile, government, space communication, and research). Frequencies within these bands are allotted to specific ITU member countries, after which individual frequencies, called channels, are then assigned to individual licensees and stations out of their respective national ITU allotments.

In the U.S. individual assignments are made by the FCC to assure equitable distribution of the radio spectrum throughout the country. Television operates on VHF channels 2 through 13 and UHF channels 14 through 83. Television uses AM frequencies for the picture and FM frequencies for the sound.

Radio waves are measured by the number of cycles per second they vibrate. (The term *radio* here is used in its generic sense, referring to any broadcast signal.) Heinrich Hertz first demonstrated and proved the existence of radio waves; hence they were referred to as Hertzian waves. Today we use an abbreviation of his name and express frequency of radio waves as Hz (pronounced Hertz) to represent cycles per second.

Television stations prefer to identify themselves by channel number, in contrast to radio stations, which choose to identify themselves by the frequency of their individual channels. What is a channel? Using the analogy of a road, visualize channels as having widths representing the amount of spectrum space they occupy without interfering with other channels. (Actually, the waves intermingle during transmission, but they get sorted out at the receiving end.) One AM radio channel occupies 10,000 Hz; FM uses a channel of 200,000 Hz; one television channel occupies 6,000,000 Hz.

Video Electronics

After a TV camera picks up a picture it changes the video into a pattern of light, then converts the light pattern into electrical impulses and electromagnetic waves. Those electrical impulses are then "piggybacked" on an electromagnetic carrier wave that broadcasts the signal through the air to be received by television sets tuned to that particular channel. The receiver converts the signal back into a pattern of light that can be seen on the television set's viewing screen.

It has been noted that in this country the visual image transmitted by a scanning process consists of 525 lines per frame and thirty such frames are broadcast every second, the selection of thirty frames being tied to our country's standard 60 Hz house current. The 60 Hz rate governs the rate of screen illumination needed to prevent flicker (called *field frequency*). The way to avoid flicker is to split the frame into two fields. This is performed by the electron gun in the TV pickup tube. The gun scans and discharges the picture information horizontally from left to right in a process called *offset* or *interlace scanning*. Thus, the electron beam scans the odd lines first, then returns to scan the even lines. The odd-numbered scan is referred to as field 1, while the even-numbered scan is called field 2. In this manner the screen will be illuminated sixty times per second, even though illumination provides only half the information of one frame. Thanks to persistence of vision our brain retains an image for a fraction of a second longer than it actually appears in real time and, therefore, we get the sense of a moving picture when actually the eye is seeing a series of lines and still pictures.

After the beam sweeps a complete field it returns to the top left of the screen to begin a new field. This return is called the *flyback* period of the blanking system. Just before, during, and after flyback the electron beam is shut off, hence it is *blanked*. This is called the *vertical blanking interval*.

Sometimes signals are sent during these blanking periods, such as closed captioning for the deaf, often using line 21 of the TV vertical blanking interval, since that portion of the screen does not ordinarily contain video information. Of course, such captioning can be seen only when decoded by special receivers in the home. Sometimes FCC-approved signals are sent during blanking periods of other lines.

TRANSMISSION

The transmitted television signal consists of two separate carriers: one modulated with video, the other with audio. The sound portion of a television signal is a standard FM wave whereas the picture is distributed by an AM wave. The two signals are sent to the transmitter by a station-to-transmitter link (STL), which may be a cable, microwave, telephone, telephone line, or a combination of any of these. The transmitter sends the signals to the antenna, which is mounted on an antenna tower. The antenna consists of radiating elements that send out waves. After the AM and FM signals reach the receiver tuned to that particular channel, the television receiver picks up both carriers simultaneously, strengthens them electronically, and accepts the picture in the tube and the audio in the loudspeaker.

The Video Signal

Essentially, the picture to be transmitted is scanned by electronic circuits that produce a voltage output proportional to the brightness or darkness of the particular area being scanned. The scanning breaks the video picture into many horizontal lines, each of which varies in brightness along its length. Basically, the picture is broken into a series of sequential lines, and a continuously varying voltage is produced, the amplitude of which is proportional to the instantaneous brightness of each point in the lines. This varying voltage is the *modulating* signal or intelligence used to amplitude-modulate a *radio frequency* (*rf*) carrier in a *multiplex* manner, as illustrated in Figure 5–1.

After transmission on one of the television broadcast channels, a varying voltage that is identical to the original modulating signal is recovered from the modulated carrier by the demodulation process. This voltage then, according to the amplitude, regulates the intensity of a beam of electrons produced in the picture tube of the TV receiver. Visual reproduction of the original picture occurs as a result of a coating on the face of the picture tube. When this coating is struck by the electron beam it emits light, with the amount of light being proportional to the intensity of the beam. Therefore, when the electron beam has a high intensity, which corresponds to a point of extreme brightness in the original scene, a large amount of light is emitted from the point on the picture tube struck by the beam. Similarly, points of low brightness will emit little light. Even without a video signal a television screen is bright when it is turned on. This brightness is called the *raster* and is produced by the electron beam being scanned across the face of the tube by circuits inside of the television receiver. Without a video signal the beam has a constant intensity, so the screen brightness is uniform across the face of the tube. As noted earlier, the scanning of the original scene at the transmitting end and of the reproduced scene at the receiving end must occur very rapidly so that the human eye sees only a full picture rather than a series of lines. In fact, the eye actually blends the lines, fields, and frames into the illusion of a moving picture just as the eye mixes together the three primary colors so that they appear to be thousands of shades.

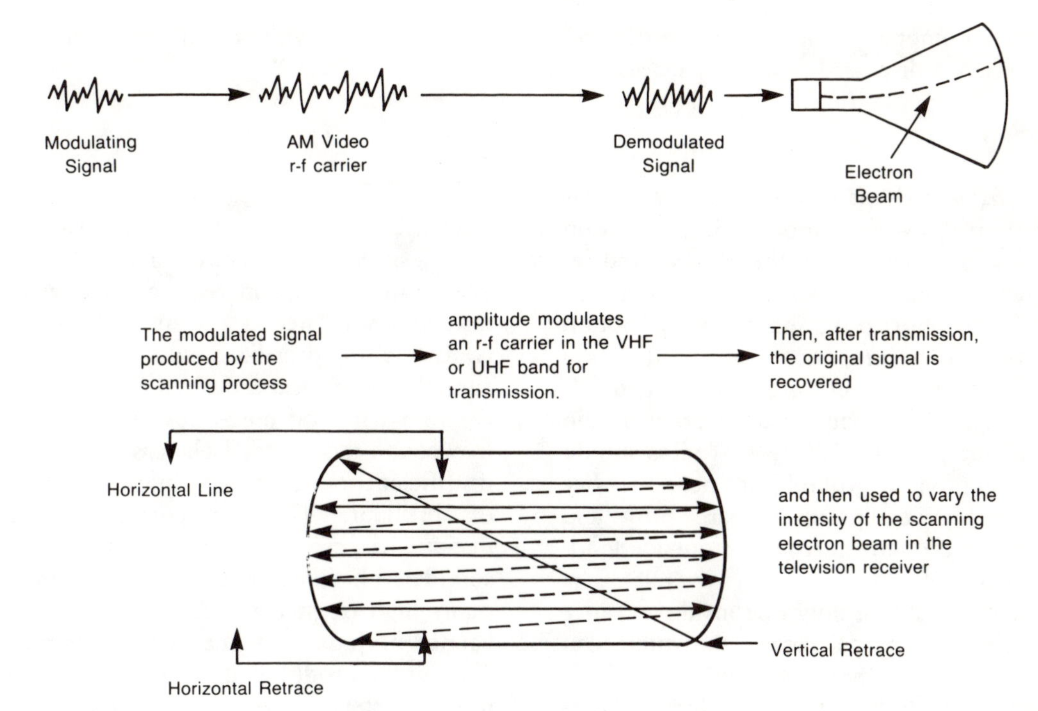

Figure 5–1 Video signal transmission process.

Color

In color, brightness components are transmitted in much the same manner as the black and white picture signal. However, color information must be transmitted by a color television signal. Color television signals are compatible with both black and white and color TV receivers. This means that a noncolor receiver produces a black and white picture of the color signal and a color receiver produces a color picture. Color is the combination of those properties of light that control brightness, hue, and saturation. *Brightness* is that characteristic of color that enables it to be placed on a scale ranging from black to white or from dark to light. *Hue* is the variable of a color described as red, yellow, blue, or green. *Saturation* refers to the extent to which a color departs from white or the neutral condition. Pale colors, or pastels, are low in saturation, while strong or vivid colors are high in saturation. Brightness is the only attribute of a color signal that can be transmitted over a monochrome or black and white system.

To produce a color image, provision must be made for the transmission of additional information pertaining to all three of the variables of color. However, because the primary-color process can be used, it is not necessary to transmit information in exactly the form expressed by the three variables. Virtually any color can be matched by the proper combination of no more than three primary colors. While other colors could have been used as primaries, red, green, and blue were selected for color television.

THE STUDIO

In the television studio are cameras, intercoms, lights, lighting boards, microphones, and teleprompters, as well as properties and sets. Each is discussed separately below.

Studio Design

The size and shape of a television studio is determined by the amount of production generated, as well as by the number and required movement of cameras and performers. Today, more programs are being shot on location; therefore, most new studios are smaller than those in longer-established stations. The dimensions of an average studio are about 30 feet × 50 feet. Studios should be located on the ground floor near a service door to allow equipment and large props and sets to be brought in from the outside loading area.

There should be only one inside entrance to a studio so that access to the studio can be monitored to prevent people from accidentally wandering onto a set during a taping or live telecast. As a rule, arrangements are made to permit visitors and talent to watch either from a special viewing producer/client booth or on monitors placed in a room where talent may also wait before going on the air.

The floor of a studio should be hard and smooth enough not to interfere with camera movements. Most floors are concrete, reinforced with "cat joints" to prevent cracking. Problems can be caused by tile or carpet on the floor; tile cracks and carpet tears. Painted concrete is most practical. In addition, there should be no columns to interfere with line-of-sight of cameras; large steel trusses rather than pillars should support the ceilings. The average ceiling is about 15 feet high to provide a way to vent the heat generated by the lights on the lighting grid, and also to keep lighting equipment out of camera view. It is essential to keep the studio air-conditioned at all times in order to keep the electronic gear in proper working condition. Lighting grids are sometimes motorized and new dimmers are controlled by a computerized lighting controller with a built-in floppy disc drive.

Acoustics

Acoustical control is critical in television studios, especially for stereo audio for television and for the increasing sensitivity of microphones. There is a science to acoustical control, with specialized equipment to measure the appropriate desired reverberation. A room with too much absorbent material may sound too "dead," whereas a room with insufficient sound absorption can be too live or "brilliant." The usual goal is to have a *live-end dead-end* (*LE-DE*) room so that sound is alive, but not overly so with too many high frequencies. Acoustical response cannot be judged by the human ear alone. The audio produced in a television studio must be listened to through the audio system for noises that a sensitive microphone may pick up or for frequencies that a particular microphone may emphasize.

Teleprompter

Teleprompting has come a long way from hand-lettered posterboard cue cards held discretely out of camera range. Teleprompters contain built-in word processing and underlining, and can handle immediate edits or inserts to keep up with last minute script changes, especially in news scripts. Some teleprompters are just below the camera lens so that the performer appears to be looking the viewer in the eye, while actually scanning the script. There are also microprompters for ENG and other hand-held cameras. Prompters for public speaking and conferences are sometimes seen in national broadcasts such as awards ceremonies or large conventions. By using glass and projection devices, the script is projected to the clear glass so that the public does not see the script.

CAMERAS

Camera Pickup Tubes

The heart of the electronic camera is the pickup tube. Some cameras need only one tube, others require more. The tubes convert the optical image into an electronic signal that then converts the light values to other electronic signals, which finally modulate a station's carrier wave by sending each picture frame separately. After being broadcast, that electronic energy is reconverted to light energy and a picture appears on the television receiver screen.

Pickup tubes vary in size as well as in the makeup of the tube's photoconductor. Outmoded now, but used for the first two decades of television following initial early use of the *iconoscope* tube, is the *image orthicon* tube, much larger than today's tubes. Use of the image orthicon began to wane when the Phillips Company developed a new tube that used lead monoxide as the photoconductor and which proved to be far superior. That tube, named a *plumbicon* after the Latin word for lead (*plumbum*), became the standard. Other tubes have been developed, including the *Saticon* from the Hitachi Company, which has given the plumbicon strong competition. The vidicon is another tube in common use. The term *vidicon* frequently connotes the lower quality tubes used in closed circuit and semiprofessional equipment, but they are used in the professional field as well. One drawback of vidicons is that they require more light than do the plumbicons.

Electronic cameras designed exclusively for studio use may have a pickup tube with a face diameter of 1.5 inches (30 mm). Others use 1 inch (25 mm), 2/3 inch (17.7 mm), or 1/2 inch. At present the 2/3 inch tubes are used in studio cameras as well as field cameras. The smaller the tube's diameter, the more difficult it is to design and manufacture. Smaller tubes, however, have led to smaller and lighter cameras.

An important development in the 1980s was the use of the *diode* gun in the tube as opposed to the older *triode* gun tube designs. Basically, the diode gun provides an improved method of focusing the electron beam on the photoconductive face plate of the tube. This results in greater resolution of the image. However, not every camera will accept a diode gun tube because the diode requires a different voltage that is not always available.

Image retention problems may result from designs inherent in certain tubes. *Lag,* which is the disposition of an image to stick on the pickup tube after the image has been removed, is caused by the photoconductor's tendency to retain an image in the same way the human eye retains a very bright image. As a result, the image lags from one frame to the next. The problem of *comet tailing* occurs when a highlight moves across the frame and leaves a trail behind it. Newer pickup tubes and cameras increasingly overcome these defects, but such problems can affect the clarity of the color they produce.

A significant technology introduced in the 1980s was the charge coupled device (CCD), far superior to tubes. Cameras using CCDs have instant start-up, high resistance to physical shock, and no burn-ins, retubing, sticking, or comet tails. Tubes continue to improve. The choice of pickup may depend on the particular use intended for a camera or on the camera operator's preference.

Types of Cameras

The U.S. television industry converted to color in the early 1970s. All cameras made for professional use are for color. Electronic cameras are classified according to the purpose for which they are designed: *electronic studio production* (ESP), *electronic field production* (EFP), and *electronic news gathering* (ENG). Each area has its own requirements.

Even equipment designed for nonbroadcast use such as industrial, educational, or semi-professional production has stringent standards suitable to the specific need. All electronic cameras share some basic similarities.

Top-of-the-line broadcast cameras use three tubes to capture the red, blue, and green components of a picture. The light of a scene comes through the lens and then is divided into red, blue, and green by a prism. Less expensive cameras may use *dichroic* filters to break up the colors. Dichroic filters, however, absorb more than one stop of light. In other words, if you shoot indoors or in low light, prism optics are essential.

The ability of an electronic camera to synchronize accurately the signals from the tubes to match the three images is called *registration*. Registration is measured with a test chart that can reveal how far the red and blue images diverge from the green. Most sophisticated electronic cameras contain some form of automatic registration adjustment to reduce the amount of time required to set up a camera at the beginning of a shoot.

Resolution in an electronic camera is mainly a function of the pickup tube and of the camera's electronics rather than the function of the lens or the internal optics. Resolving power is expressed as the number of lines that can be resolved in a given area of the image. It indicates how many horizontal lines are needed to make up the image. The best studio cameras can resolve up to 800 lines. Of the 525 lines used in the U.S., only 490 are actually used in one picture. The developr.ıent of *high-definition television* (HDTV) has demonstrated the ability to handle 1,125 scan lines.

Studio and Field Cameras

Studio cameras, mounted on tripods or dollies, are used for either live telecasts or prerecorded productions shot for later airing. Some are controlled by microprocessors. As more computer elements and chips are added, cameras will become smaller and lighter. Phillips has designed a total computer-controlled camera that automatically sets up the program, then aligns the green channel, after it aligns the red and blue to the green. The advantage of a computer-controlled camera is that it operates at 40-70% lower light levels, therefore saving power needed for lighting and cooling.

Some cameras are used both in the studio and the field. Ikegami introduced in 1986 a top quality camera for this purpose. Although weighing only 55 pounds, compared to the average studio camera of more than 100 pounds, Ikegami's HK 323 has a built-in encoder and sync generator, quality prism optics, automatic set-ups, and self-diagnostic functions. The combination camera is ideal for the smaller station that does more field work than studio work (see Figure 5–2). At this writing the 2/3 inch camera is the standard field/studio size.

ENG cameras have virtually replaced 16 mm cameras for television news location work (see Figure 5–3). Designed to be portable and easy to operate, they are also very sophisticated. Sometimes they are used by a two-person crew in conjunction with a portable videotape recorder. Generally, ENG cameras have a shoulder mount and an eyepiece for the viewfinder monitor. The main difference between an ENG camera and a EFP camera is that the ENG camera is handheld whereas the EFP camera is mounted on a tripod or dolly. Attached to the EFP camera is a studio-size viewfinder that is plugged into a remote camera control unit in order to allow necessary adjustments to camera circuits to be made by turning knobs.

The Recording Camera

Basically, the recording camera is a self-contained portable video camera designed for ENG work. (see Figure 5–4) At this writing

Figure 5–2 Ikegami 2/3 inch field/ studio camera. [Courtesy of Ikegami Electronics (U.S.), Inc.]

Figure 5–3 Ikegami HL-95 Unicam hand-held color camera. [Courtesy of Ikegami Electronics (U.S.), Inc.]

Figure 5–4 Sony BVP-1 Betacam, a self-contained portable video camera, ideal for ENG work. (Courtesy of the Sony Corporation of America.)

the 1/2 inch videotape format has superior quality to the 3/4 inch format, as discussed later in this chapter.

Recording cameras use cassettes that have two tracks of audio, plus a control track and a time code track, in addition to the picture. They are powered by rechargeable batteries mounted on the camera or by a power pack worn as a belt around the operator's waist.

A particular advantage of the recording camera is that the video can be played back immediately and viewed in the camera viewfinder. Separate playback decks can be used for transferring the cassettes to another tape format or for editing the original cassettes. In most cases the recorder portion can be removed and the camera can be used with a conventional recorder or, inversely, the recorder can be used with a different camera.

Video Lenses

Video lenses must be designed for the various sizes of video pickup tubes. The way a camera is used determines the type of lens selected. A video camera now is operated by only one person, whereas the large, early models required more people. Therefore, there is need for a single all-purpose lens instead of the interchangeable lenses found in film cameras. Almost all video lenses are *zoom* lenses, and the use of video cameras to cover sports, concerts, and other large public events has led to development of lenses with extraordinary zoom ranges.

With smaller video cameras, the all-purpose lens is achieved by means of a zoom with a set of accessories such as the *telephoto* or *wide-angle* attachment that goes on the front of the lens, or a *range extender* that mounts on the rear. Often, the extender will be built into the lens; sometimes two extenders will be mounted in tandem on the rear of a lens. Video lenses should also have a very short *minimum object distances* (MOD). The MOD is the closest point on which a lens can be focused. Close-up attachments are available to satisfy the need for close focusing without having to resort to a *macro* lens. Until the 1980s, the engineer for a video production controlled the iris remotely, or the iris was automatically controlled by the strength of the signal generated by the pickup tubes. Lenses now have an auto-iris controlled by the video signal itself, because the signal amounts to the equivalent of a sophisticated light meter.

Studio lenses are much smaller and lighter than they were a decade ago. Figure 5–5 illustrates Fujinon's 17X studio lens and shows the location of the various elements and controls. With such a lens one can preset limits

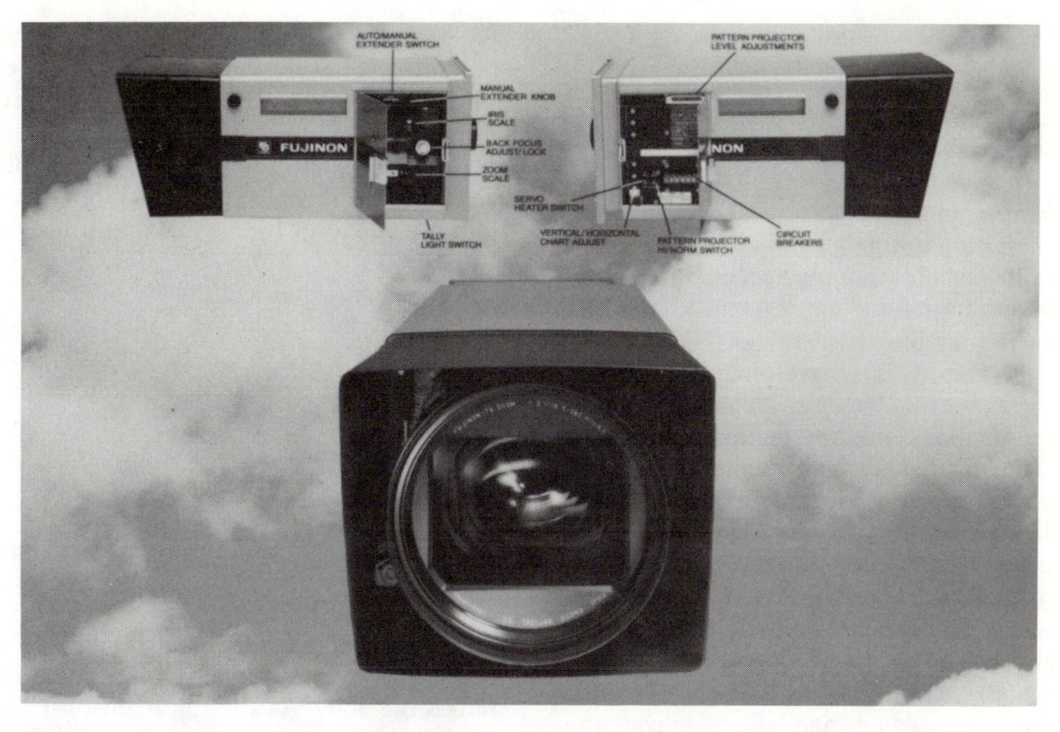

Figure 5–5 Fujinon zoom lens. (Courtesy of Gardner and Company.)

on the zoom range to match the lens to the lighting levels of every production. To change the speed of the zoom an operator uses the pan bar zoom control instead of having to reach someplace inside of the lens. There are manual overrides to enable an operator to get a particular approach to an image. This camera is much easier to adjust quickly because of easy accessibility provided by the two doors seen in the figure.

Camera developments never occur without corresponding developments in lenses. The rise of ENG has concentrated lens development on the ENG/EFP area. A major goal of modern lenses, especially for ENG cameras, is to eliminate all image vibration when shooting while walking or from a moving vehicle—whether helicopter, truck, boat, or motorcycle. A good lens can stabilize the image optically rather than having the operator depend on braces and brackets. For example, the Schwem Gyrozoom Image Stabilizer Lens zooms from 60 mm to 300 mm, enabling an operator to shoot close-ups from 1,000 feet with a perfectly steady image, yet it weighs only 6 pounds.

Computerization

Miniature computer systems are being built into camera systems, in the camera head as well as in the master control panel and the camera processing unit. The computers can complete camera set-ups, perform troubleshooting tests, and also store information about gain levels, iris, back-focus and other settings. Computers even maintain set-up parameters for separate lenses that may be required for different sporting events. The common control panels can switch cameras from stand-by to operational modes, trigger automatic procedures, and display information such as f-stop settings. The built-in computer can even adjust the gain on red, blue, and green in situations where the *Kelvin* temperature of the daylight fluctuates.

Miniaturization

Charge-coupled devices (CCDs) and computerization not only facilitate the design of smaller cameras, but may eventually replace current top-of-the-line cameras. A CCD camera may have as many as 510 elements in one chip. CCDs have already improved cameras because they provide more than 450 lines of color resolution in any format.

Miniaturized, compact cameras are used for news coverage, documentaries, commercial production, sales meetings, and seminars. The CCD microcamera by Telemetrics is only 6 inches square and 6 inches deep, designed to be mounted in tight spots either right side up or upside down. Intended for remotes, this tiny camera can be controlled from as far away as one mile to activate the zoom lens. The smallest camera in use at this writing is Optical Resources' ultramini midget camera—a wireless, 2 ounce camera no bigger than a cigarette pack. It can be fastened to the cap of a baseball player or placed inside the home plate umpire's field mask.

The design of electronic cameras is expected to change as technology changes. Cameras are already moving to self-containment and the solid state sensor is replacing the pickup tube. Also anticipated are cameras that are compatible with HDTV, which will provide a resolving power comparable to 35 mm film.

Camera Mounting Equipment

Television cameras are mounted on panning and tilting devices that permit the flexibility of control needed by camera operators. These devices are called *friction heads* or *cradles*. The cradle approximates the old rocking chair idea. No matter how the camera is tilted it remains in balance. The camera *pan* and *tilt head* are in turn mounted on *pedestals* or *dollies,* of which there are many types. The simplest is a tripod, which may be of metal or wood and which is usually mounted on a three-wheeled dolly. The principal disadvantage of this type is the lack of vertical movement except by manual adjustment of the tripod itself. More sophisticated pedestals incorporate mechanisms for readily raising or lowering the camera. Some use a hand wheel. Others are so carefully balanced by weights that the pressure of a finger on a large ring is enough to move the heavy camera up and down. Others are activated by electronic motors that operate very quietly. Still others are operated by compressed air.

In larger studios the use of power-operated dollies is common. These electrically controlled devices enable the camera operator to maintain complete control of the camera while a second person mechanically moves the dolly. As a result, many obstacles are avoided. Camera cranes, similar to those used in motion picture production, are also found in large studios. The camera operator rides the crane in a seat and operates the camera conventionally. Gross movements are controlled by dolly manipulators who move the camera boom in any desired direction. Such cranes easily elevate the camera to 10 feet or more and can lower it almost to the floor of the studio.

An example of the late 1980s state-of-the-art is Vinten's remote servo camera control system, which provides remote control of pan, tilt, zoom, focus, iris, and pedestal height. Sophisticated systems of this nature enable some news studios to be so completely automated that one news reporter can operate all the equipment while presenting the news on the air. In Figure 5–6 are examples of a pedestal, a tripod and camera head, and a crane.

Not only economics, but studio space determines how elaborate a mounting can be usefully employed under average production conditions. Usually, the more complicated or cramped the settings or studio, the less op-

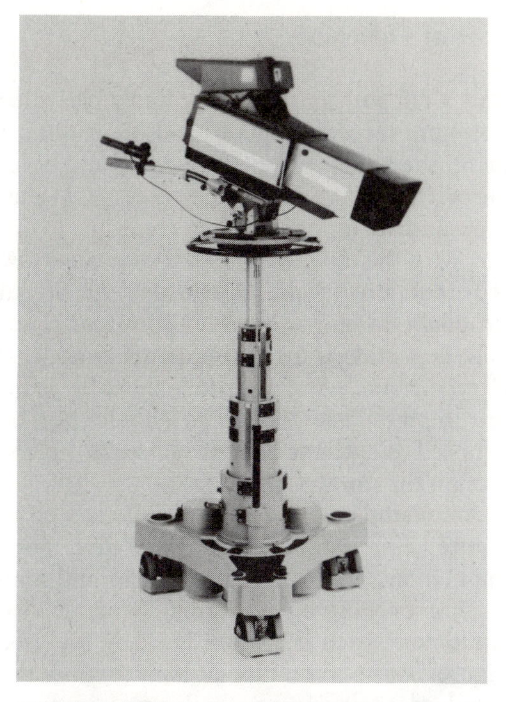

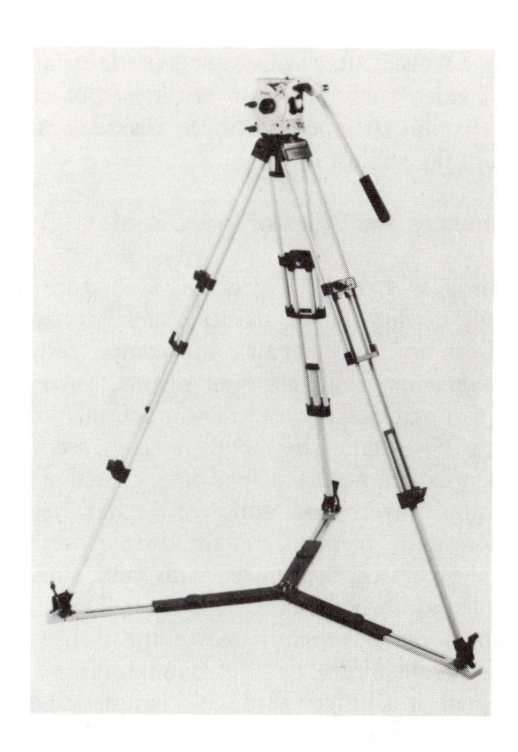

Figure 5–6 Some components for remote camera control systems. Hawk extended range pneumatic pedestal *(top left),* Vision 10 fluid pan and tilt head with Vision two-stage ENG tripod and Vision heavy duty spreader in full elevation *(top right),* and Merlin camera arm on teal standard range pneumatic pedestal and crane *(bottom).* (Courtesy of Vinten Equipment Inc.)

portunity to use cranes. For field work at sporting events and similar remote pickups, tripods with or without dollies may be used when actual movement of the camera from one position to another is not needed.

VIDEOTAPE

Videotape quality has come a long way since Ampex introduced the first videotape recorder (VTR) in 1956 using 2 inch tape on large reels. Today videotape is used not only for video cartridges and cassettes, but also for computer tape and floppy disks, as well as in the studio.

Structure and Types of Videotapes

Through a complex process, a mixture of high quality metal oxides (such as iron, chromium, and cobalt), lubricants, resins, dispersants, head cleaning agents, solvent, and fungicides is coated on a special polyester base film. The metal particles used in the mix are needle-shaped and coated with various layers and adhesives. After being thoroughly blended, the mixture is dried, heat-processed, then milled with sand, stones, and smooth balls. Next it is filtered and remixed under pressure. After the polyester tape (called *film* by the manufacturers) is coated, it is moved through a magnetic field in order to position the needle-like magnetic particles in a longitudinal direction in which they must not touch or interfere with each other. If the mixture is spread unevenly on the film base it can cause dropouts to occur during shooting. Quality tape is essential for professional work. Newer generations of videotapes will require even smoother surfaces and thinner film, using pure metallic particles or thin metal films in order to achieve longer recording times and better signal-to-noise ratio. Recording density and capability has been increased approximately one hundred times since the four-head broadcasting VTRs were first introduced in 1956. Future tapes will increase density at least another ten times and will be applied to the film surface by electron beam or other still undeveloped techniques.

Two-inch tape is disappearing. Broadcast quality tape has moved from 2 inches to 1 inch, 3/4 inch, 1/2 inch, and even to 1/4 inch. An 8 mm metal tape and an 8 mm recording format may soon become the standard for ENG. New tape will have to be designed to abide by the SMPTE standardized 19 mm format required for digital VTR's.

Care of Videotape

The VTR path must be absolutely clean for optimum recording quality. Also clean must be all the audio, video, time code, and erase heads, in addition to all tape guides and other tape path components that come into contact with the tape. Dirty heads can cause low reproduction levels, distortion, or signal dropouts on one or more VTR output channels. In addition, the videotape recorder used to record and wind videotape must be properly maintained. Videotape should not be exposed to extreme heat or moisture (see the section on storage, below).

An example of poor tape image is seen in Figure 5–7, which illustrates that poor maintenance can cause poor video and may require expensive reshooting. Maintenance equipment, such as "TapeChek" by RTI (see Figure 5–8) plays a tape at twenty times the standard speeds and evaluates, inspects, cleans and polishes both clean and recorded tapes as well as identifies dropouts. Thus, recorded tapes can be checked without being erased. This equipment also "vacuum cleans" the system by reducing temporary dropouts caused by loose oxide and dirt.

The color picture is stored on a single coating of magnetic particles that does not change with time. Because the magnetic strength (*coercivity*) of videotape has doubled in recent years, the stronger magnetics make it very difficult to erase the tape except with specially designed erase heads containing a specially designed videotape demagnetizer. For security, however, it is important to keep a tape at least 1/2 inch from any large magnetic field, such as a motor or a transformer, so that there is no possibility that the tape might be erased accidentally. If you need to transport tape by air it is helpful to know that X-rays cannot affect magnetics; however, the hand-held magnetometers used for a body search could erase tape. In other words, do not carry tapes on your person when you pass through an airport check.

Videotape is not affected by light; but heat

Figure 5–7 Example of poor image resulting from improper videotape maintenance. (Courtesy of RTI.)

and moisture can cause deterioration by causing the binder to break down. Short-term exposure to high temperatures or immersion in water cause little noticeable degradation, but a tape exposed to such conditions should be duplicated if long-term stability is necessary.

Storage of Videotape

If studio tapes are allowed to go ten years before being used they should first be played through to the end at normal speed. It is important to avoid high speed on the first pass because of the possibility of the pack's slipping and causing fold creases in the tape.

A properly stored videotape of archival value should last for hundreds of years if it is protected and stored properly. It should be stored in an upright position in a temperature-humidity controlled room with humidity ranging between 34 and 45°F and the temperature in the low 60s (°F). Each reel should be wound on a special tape winder before storage and once again when removed in order to equalize the tension stress in the tape pack. Every three to ten years each reel should be wound end-to-end to relieve built-in stresses. Ideally, the tape should be wound in the same environment it is to be stored in. A day or two before storage the tape should be placed in a dry environment after which it should be sealed in a quick-seal plastic bag. The prepared tape should be stored on steel shelves in a room with no combustibles.

VIDEOTAPE RECORDERS

Until 1956 film was the standard medium for recording sound and visual images. Ampex's introduction of the first major VTR started the videotape revolution. That first VTR was a bulky machine that took up most of the floor space of a small 8 × 10 room.

Videotape recorders are usually identified by the width of the tape they use. The first tapes were 2 inches, followed later by the 1 inch, used on a reel-to-reel machine. Then came the smaller forms in cartridges and cassettes: 3/4 inch, 1/2 inch, and 1/4 inch.

In 1967 the first portable, record-only VTR appeared on the market, developed for industrial and institutional use. Initially it did not have broadcast quality, but the call for

Figure 5–8 RTI's TapeChek Model 6120. The device evaluates, polishes, and cleans blank and recorded 1 inch videotape. (Courtesy of RTI.)

a lightweight portable video camera to serve electronic newsgathering resulted in a number of manufacturers developing videotape recorders for broadcast professionals to use in the field and for editing news stories. Today's VTR is often part of a recording camera that is a self-contained video camera with a recorder, called a *camcorder*, capable of recording broadcast quality video on a 1/2 inch cassette (see Figure 5–4).

Tape Speeds

VTRs use one or more of the standard speeds, comparable to the speeds of audio tape: 15/16 inch, 1 7/8 inches, 3 3/4 inches, 7 1/2 inches, and 15 inches. The figures represent the number of *inches per second* (ips) that pass the recording/playback head. Speeds of 7 1/2 ips and 15 ips are preferred because they provide better audio, picking up the widest range of frequencies at higher speeds.

Wave Form Monitor

An important piece of equipment is the *wave form monitor,* which allows the television system technician to monitor video signals and to determine when adjustments are needed. The wave form monitor also shows the various shades of color and of black and white, indicating whether the shades are evenly spaced. It further allows the technician to balance cameras so that they will generate pictures of similar brightness and color.

Video Cart Machines

Before the development of the cartridge or cassette machines, short program segments such as commercials and public service announcements had to be copied onto a master tape that was then edited into the videotaped program material. That caused extra wear on the recorder heads, tied up the VTRs for *dubbing* (making duplicates), and lowered the signal quality of the copied then recopied short segments. Normally, each new generation of a dub may deteriorate. That led to the development of cartridges, each of which held only one short segment or spot. A *cart* is a specialized plastic container that houses a short, set length of videotape. Carts are stacked in racks and automatically cued to start on signal, then manually returned to their places in the rack. Some systems can handle more than one thousand events in any tape format or combination of formats. A 1980s-developed cart machine by Odetics holds as many as 280 carts on-line and can track sixty-five thousand carts in the database.

Videocassette Recorders

As previously mentioned, the 3/4 inch, 1/2 inch, and 1/4 inch tapes are housed in *cassettes;* these have different and smaller plastic housings than do video cartridges. There are two major formats, Beta and VHS, which also differ in size. Note that the home videocassette recorder (VCR), whether VHS or Beta, is different than the 1/2 inch professional recorder, most noticeably in the tape transport speed, and does not have the component video recording system of the professional cassette.

Videodiscs

In 1977, although many firms were working on designing videodiscs, Phillips-MCA and RCA took the lead in production and distribution. The *videodisc* is a phonograph-type system capable of playing color video programs. The videodisc can scan fast forward or reverse almost instantly to a particular frame, yet is never touched by anything but a laser optical beam; thus it is never scratched and will always provide top fidelity. The videodisc requires a special playback unit.

VIDEO RECORDING FORMATS

There are a number of video formats from which a producer may choose. Each has its own particular advantages and disadvantages. Video formats can be identified according to (1) the type of signal into which the visual image has been translated, (2) the way in which the signal is recorded, (3) the physical characteristics of the medium onto which it is recorded, and (4) to some extent the size of the tape used for recording the video signal. However, the physical size is not an important limiting factor; the limitations are more a function of the speed with which the tape moves across the recording head (i.e., the *bandwidth* or range of frequencies which can be recorded).

When we consider the way the signal is recorded, we must consider how the tape and recording head move in relation to each other. Formats in use at this writing include 2 inch quadruplex, 1 inch helical type C, 1 inch helical type B, 3/4 inch heterodyne, and 1/4 inch (which uses a CCD). Two inch and 1 inch tapes are reel-to-reel, 3/4 inch and 1/2 inch are used in video cart or cassette, while 1/4 inch is on videocassette.

Quad

The earliest tape format was the 2 inch with four recording heads on a drum, rotating at 90 degrees to the direction of the tape travel so that the track records across the width of the tape as it moves past the drum. Program audio is on one edge while an audio cue track and a control track are on the other edge. The quad tape moves through the machine at only 15 ips but can scan at the equivalent of 1,560 ips. *Luminance* (black and white) and *chrominance* (color) are recorded together on one encoded signal, not only on quad but also on 1 inch-C and 1 inch-B. While quad used to be the best format it is now being gradually replaced by 1 inch, which is of equal quality and is not as bulky. The 2 inch quad recorder designed for production or postproduction is more than 5 feet tall and weighs 1,100 pounds. Its advantage is that it can record 250 minutes on one reel.

One Inch Helical Type C

One inch type C format grew after a standard was adopted because the format was less expensive, less complex, and offered better pictures than the 2 inch. This format also provides slow motion, electronic slides, and moviola-type editing with the picture visible in the search mode. One inch videotape recorders have a helical configuration in which the tape is wrapped around a drum containing a rotating head placed at a shallow angle to the general direction the tape is traveling. The track is recorded on the tape as a series of sloping adjacent tracks, each of which contains a complete field of the video image. In type C the tape moves through the machine at a rate of 9.61 ips while the head moves across the tape at a rate of 1,007 ips. In the late 1980s this was a preferred production format (see Figure 5–9).

One Inch Helical Type B

In type B format there are two video heads moving across the tape, just as in the quad machine. One of these heads is always in contact with the videotape, allowing a continuous video signal to move at 9.65 ips through the machine while the video head writes at a speed of 945 ips.

One Inch Capabilities

One inch VTRs have advanced to the point where they even have self-diagnostic capabilities. For example, in either record or playback mode the Hitachi HR-230 1 inch contains a computer inside to monitor forty-nine performance parameters second by second. The information in its memory can be recalled in words as well as in numbers and symbols. It can also provide real-time reverse and field/frame still motion.

Three-Quarter Inch Heterodyne

Standard 3/4 inch cassette video recorders have a much slower tape speed of 3.75 ips and a scanning speed of 404 ips. The slower tape speed requires a different method of processing the color elements of the signal. In fact, the word heterodyne means "color under."

Because of limitations of the bandwidth, black and white and color cannot be recorded directly as in the 2 inch and 1 inch; instead, the two signals of lesser bandwidth are recorded separately. The major deficiency is that each time a tape is dubbed, the later generations produce an increased delay of color and black and white, which sometimes causes a herringbone effect. Heterodyne 3/4 inch is used mainly for production for ENG or for *off-line* editing. When this format is used, it is frequently transferred to 1 inch for postproduction editing. Preliminary editing is often done with 3/4 inch cop-

Figure 5–9 One inch type C helical videotape recording format.

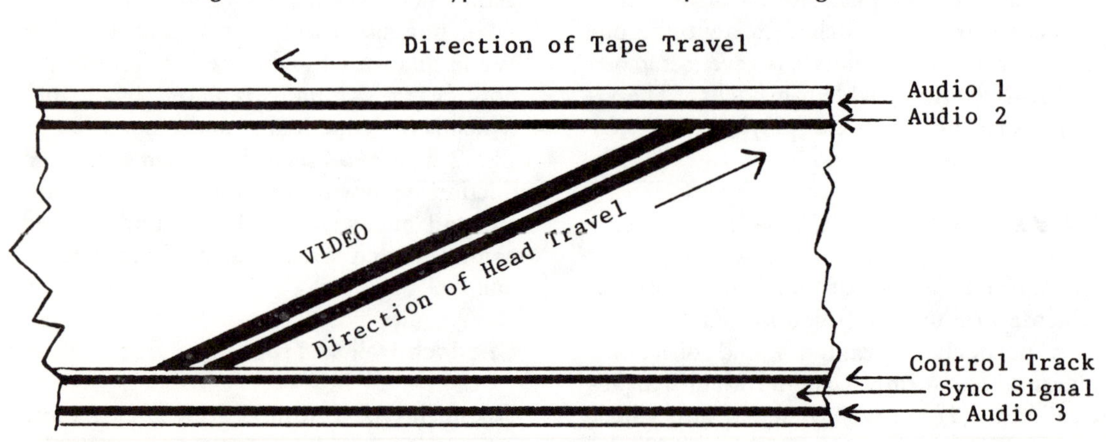

ies of the original tape because of the difference in cost and the ease of operation between the two. It is also used as a distribution format for productions shot in 1 inch or 2 inch.

The 3/4 inch format is built into a portable editing system by Camera Mart, in which the editor is supported by two time-base correctors, plus time code, dual floppy disk drives, a high-speed printer, a switcher, and an audio mixer. In addition, Camera Mart's system includes two color monitors, a wave form monitor, vectorscope data display monitor, and a two-channel audio system. The Sony U-Matic is particularly good for component processing. In 1986 the JVC Company introduced a 3/4 inch editing recorder/player.

Half Inch Format

In the mid-1980s the 1/2 inch format became surprisingly successful. Presented in both VHS and Beta forms, 1/2 inch surpassed the quality image obtainable with 3/4 inch videotape recorders. Built into a portable recording camera, each cassette can contain twenty minutes of broadcast quality video and is used extensively in ENG and other field work. The video contains two tracks of audio plus a control track and a time code track.

It is possible to record a broadcast quality signal on 1/2 inch and 1/4 inch cassettes because of three factors: (1) tape speed is substantially increased compared to the speed of home cassette recorders; (2) recording density in the new formats is more than twice that of any other videotape recording format; and (3) the luminance and chrominance of the video signal (i.e., the brightness and color) are recorded separately. The increased tape speed improves the audio quality as well.

M-Format—Chroma Trak

One of the most important 1/2 inch developments is the format designed by Matsushita-Panasonic in conjunction with RCA. Matsushita calls it M-format while RCA has named it Chroma Trak. The original M-format has been replaced by M-II, which is not compatible with the original M-format system. A particular advantage of M-II is that it provides the long-awaited long-playing time, achieving a full sixty minutes using standard half inch VHS cassettes. Using metal particle tape instead of oxides, M-II has broadcast image quality thanks to its 5 MHz bandwidth. Its stereo audio quality is also excellent. The image produced by the M-II format is superior to that of the best 3/4 inch video recorders; thus the original image is sharper and clearer and holds up much better as it is dubbed through several generations. Another advantage is that the time code, audio, and control tracks are recorded on longitudinal tracks on the edges of the tape in such a way that the video can be edited without interfering with any of the other tracks. M-II has been submitted to the SMPTE to consider for adoption as an industry standard.

Sony's Chroma Trak, using Beta cassettes, is another 1/2 inch metal particle 1/2 inch format. In this form, twenty minutes of material can be recorded on an L-500 Beta cassette, which is smaller in external dimensions than a VHS cassette. Sony is working to extend Beta's M-II playing time to thirty minutes, which is considered essential for automated on-air operations. The growth of 1/2 inch has increased industry interest in 1/2 inch component systems, discussed later in this chapter.

Quarter Inch Format

Using *compact videocassettes* (CVCs), which are slightly larger than an audio cassette and one-third the size of a Beta cassette, the 1/4 inch format uses a CCD to temporarily store each line of video prior to recording. Using helicon scan, it records each line of video on a parallel track. The net result is that the Bosch 1/4 inch format can record broadcast quality video superior to the signal recorded by a 3/4 inch video recorder; in addition, it has two audio tracks and a time code track. The 1/4 inch format can record twenty minutes of material at a tape speed of 5 ips. Hitachi and Ikegami have also entered the 1/4 inch market. However, observers predict an uphill battle for 1/4 inch video in the face of the 1/2 inch format's increased acceptance.

Future

Digital videotape appears to be satisfactory for playing back commercials and might be seen in future production of videotape recorders. Introduced in 1986 and considered excellent for spot playback is the multiple event record and playback system (MERPS). While 8 mm has proven its value for industrial formats and other nonbroadcast applications, it is not expected to be in general use in broadcasting until at least the 1990s. All three networks have already changed to 1 inch type C video recorders, although some experts think the type B is better. Network choices, of course, influence the choices of individual stations, independents as well as affiliates. Most engineers, however, believe that the differences between the two recorder types are small.

DIGITAL TELEVISION TAPE RECORDING

Originally, all audio and video was recorded by *analog* signal, but in recent years *digital* technology has grown in audio and video. What are the basic differences between an analog signal and a digital signal?

An analog signal is a continuous electronic wave that imitates the shape of the sound wave. The video signal generated by the pickup tube is also a continuously varying signal in which the amplitude corresponds to the brightness level of the image on a particular scan line. All video signals, just as all audio signals, originate as analog signals; but it is possible to convert them into digital signals. The term digital refers to a method of converting information into an electrical signal by translating information into *bits* (a word condensed from *binary digit*).

The main advantage of a digital signal is that it is not subject to distortion or degradation, unlike an analog signal, when it is processor reproduced. There is likely to be an increase in *noise* every time an analog signal is rerecorded or processed. Video noise is seen as streaking, snow, color problems, or luminance noise in recorded signals. In addition, digital information can be stored in memory to make random access possible. For example, a computer using digital standard conversion equipment has already increased U.S. foreign distribution of programs on videotape by converting the U.S. signal to signals compatible with those of other nations. Digital manipulation of an image also makes possible a variety of image enhancement techniques used to make tape-to-film transfers.

Digital can also be used for special effects with special equipment and digital graphic generators. For example, images can be reduced to fill only one small part of the screen, or can create a zoom on one frame of video. Digital can combine frames into multi-images, reduce frames and rotate them anywhere on the screen, and achieve flips, wipes, freeze-frames, and strobe effects. Digital can also modify color, brightness, contrast, and polarity, all by pushing buttons instead of using cameras.

Because no manufacturer had committed itself to a particular digital VTR design, standardization was possible for digital: a single component standard for 525-line and 625-line systems. A standard does not dictate any particular recorder design, but establishes the specifications of the signal to be laid down. The goal is a truly international standard. Executive vice president of operations and technical services for NBC, Michael Sherlock, predicts that by 1993 NBC will have a plant dominated by component digital technology.

Although at this writing tape is still dominant, there are other video recording systems gaining acceptance. The laser disc recording system can convert videos from NTSC to PAL. Hard computer disk applications, such as Abekas', are oriented to animation and videodisc mastering more than for direct broadcasting.

COMPOSITE AND COMPONENT VIDEO

Composite Video

Prior to the early 1980s there was only *composite* video. The television signal that comes out of all conventional TV cameras is the same signal that is broadcast to home receivers. This is called *composite* because all elements of the TV picture are included in this one fluctuating voltage, called the *signal.*

The first television signals were black and white. When the FCC made station allocations it determined spacing between station signals according to the bandwidth (frequency) necessary for a black and white composite signal, which consisted of luminance (light and dark information making up the black and white horizontal picture), horizontal sync pulses needed to trigger the horizontal scanning circuits, vertical sync pulses, closed captioning, and other information. The composite video signal is made from just one analog signal which contains all that information.

The composite *color* video signal is made from the analog signal after the color has been encoded onto the luminance signal. The black and white bandwidths did not leave enough room for the necessary large amount of color information, but engineers created a sophisticated method of color encoding and the NTSC designed a color system compatible with black and white sets. Color information is based on the primary colors of red, green, and blue. That means three color signals are needed. Black and white is the sum of all three color signals and engineers recreate one color by algebraically subtracting two colors from the black and white signal. This is the NTSC system. As it is broadcast on the carrier and subcarrier of the television station, the color NTSC signal enters a TV set or a monitor and, with special filter circuits, returns the single encoded signal into its three original parts. The NTSC system works satisfactorily for live television, but there are shortcomings when recording, storing, and retrieving an NTSC composite signal from a videotape.

Component Video

Because the composite color problem affects only recorded video, engineers created a different recording process that uses more space during the recording and editing process, and then compresses the final signal back to NTSC before it is broadcast. The principle of *component* recording is simple: it does not mix black and white and color signals before recording. Component systems use two separate recording channels, one video channel for the black and white luminance signal and the other video channel for the color information. By keeping these separate there is no overlapping or degradation caused by separate filter circuits. (Figure 5–10 illustrates a component video mixer.)

As this is written there are not yet standards for component recording. Each manufacturer uses a different system. A tape from the Beta system is physically and electronically incompatible with a VHS-M format. Because of its two-channel nature, component is not compatible with conventional equipment and cannot be edited on contemporary machines; nor can component be switched or routed in conventional studios and, therefore, it cannot be broadcast as is. However, because all component VTRs have an NTSC output that does interface with all existing systems, the component signal is

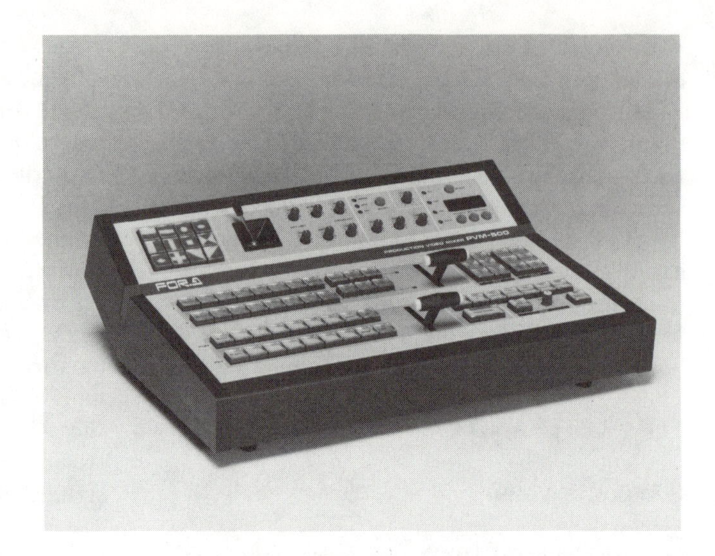

Figure 5–10 The FOR-A PVM-500 component video mixer for post-production component processing. (Courtesy of FOR-A Corporation of America.)

converted to NTSC, enabling the component video to be broadcast.

Component switchers, signal processors, and monitors that enable pure component-to-component editing are other elements in the component system. At this writing component digital is used primarily for graphics and postproduction work. It is expected that the industry will continue to work toward the goal of digital component video.

VIDEO POSTPRODUCTION

Originally all television was live and the only means of recording the images was by kinescopes; that is, by filming the picture directly off the receiver. Electronic technology and video production now provide distinct control over all postproduction phases, including audio. Editing is the key to postproduction because it adds punch and timing to video recorded material. Editing videotape is more difficult than editing film because in film you can see each frame, whereas with videotape the visual information cannot be seen without electronic assistance. (Figure 5–11 shows a late 1980s state-of-the-art video editing system.)

In the late 1950s videotape editing was a cut-and-splice process that used essentially the same method as for hand-splicing audio tape. It was a complicated, slow, and tedious job—and imprecise. In the early 1960s electronic splicing systems were introduced that allowed transfer of recorded material from one VTR to another under precise controls that ensured perfect frame-to-frame match-up at the splice. However, one still had to find the desired splice point by trial and error, then to mark that spot with electronic tones or pulses for later editing. Thanks to the development of computer technology and microprocessor chips, Ecco Inc. designed the first practical time code editing system and introduced it in 1967 by creating the SMPTE time code applied to the tape by a *time base corrector* (TBC).

Time Base Correctors

The SMPTE/EBU time code, recorded by various methods, provides each video frame with its own identification indicating the hour, minute, second, and frame number. The binary pulse-coded electronic signal is recorded either on a separate track area or on a portion of the video signal that is not

Figure 5–11 Ampex VPR-6 with overhead bridge tape console, a type C 1 inch computerized video recording editing system. (Courtesy of Ampex Corporation.)

used for picture information. After the SMPTE approved of Ecco's "On Time" original system, that standard was adopted by the EBU, thus making it a truly international standard that is used as the basis for all current sophisticated video and audio tape editing in synchronous systems.

Digital time code uses binary numbering systems (0, 1, 10, 11, etc.) because digital editing is faster and more accurate. The binary system permits computers to interface with editing terminals to create fully automatic edits. Not only is time code precise, it also provides interchangeability between editing systems and allows precision synchronization of one VTR to another, matching frames perfectly with its "electronic sprocket holes."

TBCs can now handle all sizes of videotape. Auxiliary equipment that accompanies time code readers includes time code generators, time code reader/generators, time code analyzers to analyze SMPTE time code for a series of diagnostic faults, time code inserters, and transport synchronizers, in addition to digital clocks and timers. To make full use of the time code it is wise to keep a time code–indexed log while the recorded program is being shot. In this way the editor can determine in advance what is wanted and can go right to a selected point on a tape simply by entering a time code index number into the editing system. It is important to maintain proper levels when recording because audio amplifiers used to distribute the time code may be overloaded if excessive levels are fed into them, in which case time code distortion and erratic time code reading will result.

Editing

Two types of edits are possible with the SMPTE/EBU time code system: *assemble editing* and *insert editing*.

Assemble Editing

Assemble editing involves adding new recorded material to the end of previously recorded material. In assemble editing the video and/or audio material is transferred from the source VTR to the master record VTR, which instantly starts laying down new control track pulses at the start of each new edit. Assemble editing requires use of an appropriate slave generator that resynchronizes the new time code at the start of each successive assemble edit.

Insert Editing

Insert editing involves insertion of new recorded material into previously recorded material and lays down the new material over existing video and audio. With insert editing the control track signal already on the edited master tape remains untouched.

Off-Line and On-Line Editing

Video postproduction relies heavily upon *on-line* and *off-line* editing. In general, such editing takes place in an editing suite that can include digital special effects, a title generator, a color corrector, a multichannel audio mixing board, and a computer with both a paper punch and a high-speed printer. An edit controller can also be connected to a floppy disk storage system as well as to VTRs.

Off-Line Editing

Off-line editing in postproduction occurs before on-line editing and refers to editing that does not directly edit the master tape. Time code makes it possible to interchange among editing systems and to save money and time with a process called autoassembly. Programs can thus be mastered on high quality VTRs, then dubbed down with the same time code to a less expensive tape format (usually a 3/4 inch cassette), which is then edited. Once the off-line editing is completed, the master tapes are edited during an on-line session. If a computerized off-line system is used, the on-line editing may be a completely automated process.

When a 3/4 inch dub is made for editing purposes it is possible to display the time code in the picture signal for each frame. Such copy is often called a *window dub*. Window dubs can frequently facilitate the editing process but cannot be used if the tape is intended for broadcast. Some stations "burn" the time code into the off-line tapes for quick reference.

On-Line Editing

The result of the off-line editing session is fed to an editing system that controls high quality VTRs on which the master tapes have been mounted. This is called an on-line editing system. The time code in the off-line edit list corresponds directly to the time code on the master tapes, and all the on-line edits are performed as fast as the VTRs can run the tape to the designated edit locations. Therefore, expensive running time on the mastering VTRs is kept to a minimum. In addition, the valuable master tapes are never touched during the edit decision-making process and are therefore protected.

Editing of the sound effects and nonsync music is generally done after the master tapes have been conformed to the edited tape. Optical effects can also be performed electronically during the on-line session, including fades, dissolves, wipes, freeze-frames, step printing, frame blowups and reductions, superimpositions, and traveling mattes. Scene-to-scene color corrections are also made, if needed, during on-line editing.

It is possible, of course, to perform all editing during the on-line phase. The main drawback is that on-line editing facilities and personnel cost much more than do off-line; it can be expensive to experiment during on-line editing. A paper edit can be used to cut the cost of the on-line editing, while incurring minimal costs in off-line editing by using a 1/2 inch cassette on some Betamax or VHS format machines.

DIGITAL VIDEOGRAPHICS

The arrival of *digital videographics,* formerly called component videographics, was created by the marriage of the character generator to computer graphics, and has provided increased opportunities for creative video, special effects, animation, still store, and the paint box.

The Character Generator

Before the development of the electronic *character generator* in the 1970s, written or graphic information such as program titles, credits, sport scores, and weather information were printed on poster boards in the studio's art department. The cards were then placed on a rack in the studio for a camera to shoot as needed.

From the mid-1970s to the mid-1980s character generators evolved from simple, low-resolution, dot-matrix monochrome titling systems to high-resolution multifunction graphics systems. By the mid-1980s the lines between graphics and character generators began to blur. Most sophisticated video editing facilities now include a character generator with a great variety of type faces that can be generated in any number or size and positioned anywhere on the screen. The letters can be colored and given a drop shadow. The information is typed through a keyboard and is stored in the generator's memory bank. Such generators can also "draw" on the screen, using an electronic stylus with a digitizing pad, or can color video images. When information is converted to digital form, the generator may be interfaced with a larger, more complex computer.

Graphics

Top-of-the-line character generators rival the capabilities of special effects generators. The joystick makes it possible to shuttle the videotape backwards and forwards and to move through thousands of colors. For example, systems such as Chyron IV include diagonal typing and multicolored logos. A single graphic can display as many as seven colors, which can be recalled easily with a single keystroke and positioned anywhere on the screen. Background graphics can be created quickly with either a camera or a digitizing tablet and then displayed as a still-store with a high-resolution text, including three-dimensional animation and weather images from a choice of weather services.

In 1985 ColorGraphics' ArtStar II introduced a color breakthrough with a graphics production system that offers sixteen million color choices. Character generators also enable production houses as well as broadcasters to handle in-house messages and master control inserts, keep signals and tapes sorted, as well as insert tape IDs, time, logos, titles, slate boards, indexes, and time code.

Animation

Animation has advanced tremendously with digital videographics. In 1986 animation was the buzzword among broadcasters. Bosch introduced what it calls a "truly real-time graphics animation tool" that, in real-time, builds objects and establishes animation key frames and sequences. Another development in the 1980s were systems that allowed random access recording and playback of video frames, enabling the editor to make an internal edit or animation frame playlist without ever once actually editing the tape.

Electronic Still Store

Television studios keep thousands of slides and still photographs on file for background sets, graphics, or reference use. Easy retrieval has been difficult: providing space for storage, not being able to find desired slides or photos, and getting scratches and dust on the stills. In the late 1970s Ampex marketed the Electronic Still Store, which records, stores, and plays back still pictures for television broadcast. The system uses digital techniques to store video images on a computer-disc memory and assigns an address to each still to facilitate rapid and accurate access. Access to any one picture can be obtained in less than one-tenth of a second. The system has two simultaneous and independent outputs, important when dissolves to another still or preview are required. Selection may be preprogrammed for sequential delivery.

Digital Disk Recorder

At the heart of some character generators is a dual microfloppy disk drive system for quick access to character font and page information. This is the type of equipment that can be used for teletext. Digital disk recorders with storage capacities are also used for single-frame editing applications, including mastering videodiscs, animation, and complex multilevel matte work.

TELECINE

The word *telecine* is a combination of the words *television* and *cinema.* It refers to the airing of film and slides as part of programs, commercials, spots, or station IDs. Simply, telecine is a television camera that happens to take its input from a film projector or slide chain. There usually is a separate telecine room housing a number of slide projection systems and film projectors.

Slides

Through telecine, slides can be fed into a telecine camera for either production or live telecasting. Technicians set up the slides in appropriate carousels that are later activated upon demand either by a switch in master control or by a computer-controlled system that can handle some of the feeds automatically. Television uses 35 mm slides for many visuals. A slide chain involves two carousels or drums that air alternately to avoid empty air as one slide is changed to the next. A complex arrangement of mirrors at a 45 degree angle feeds the second slide's visual into the telecine camera.

Film

Film in telecine works much the same as do slides. Television stations use either 16 mm or 35 mm film; the projector used is much like that found in the home, modified somewhat for television. The main difference is that for television the image is focused on the camera's pickup tubes rather than on a screen. Film projection, like slides, requires mirrors for changing reels.

Film is seldom used for direct broadcast, thanks to the development of videotape and videographics. It is necessary, however, for a station to maintain equipment to handle film because material arriving in film format is normally transferred to videotape in telecine before being edited or aired. Film is used more by smaller stations that cannot afford sophisticated transfer equipment and that receive many programs in film. Sometimes negative film is transferred to one inch videotape for on-line editing.

Frame Rates

A film projector normally runs at twenty-four frames per second, whereas television operates at sixty fields per second. During airing or transfer to videotape, these rates must be made equal for electronic reasons. The projection telecine overcomes the problem by projecting the first frame twice, the next frame three times, the next twice, the next three times, and so on. Thus each film frame is projected an average of 2.5 times. Multiply 2.5 by 24 and you get 60, which equals the television field rate. The frame rate can be varied when using the flying spot scanner or a CCD system. State-of-the-art electronics permit the frame transport rate to be varied continuously from zero to forty-eight frames per second.

Flying Spot

The *flying spot* is a flying electron beam from the *cathode ray tube* (CRT) that is produced by synchronization of the sweep of the CRT with the film transport. The flying spot provides better resolution, better signal-to-noise ratio, and better color than the projection telecine, and is best for transfer work. For direct broadcast, however, the projection telecine is preferred because the technology is the same as the cameras used at the television station.

Control

Telecine is controlled either by switches from the master control room or by computer. As digital videographics grow, and film is used less on the air, the size of the telecine room is shrinking. Computers can start and stop the projector and/or slide system as well as turn the equipment on and off, or automatically change slides or films. Some stations have retained only one film projector in telecine, solely for the purpose of transferring film to other formats.

State of the Art

Telecine equipment of the 1980s uses digital processing and CCD image sensors to produce high-quality video pictures. There is no registration drift and no tubes to go bad, and digital telecine interfaces with all available color correctors and can convert into all international standards. The trend is to upgrade telecine using CCDs. Another new system features a magnetic head for recording time code on film stock. For high-end graphics applications, a new antiweave gate steadies the film for better control. The *Lokbox* permits film sound editing techniques to be used with the video picture by means of a synchronizer that synchronizes a video machine to film transport. A further recent development is HDTV, with its superior video resolution.

THE CONTROL ROOM

The studio control room is the brain cell responsible for directing all studio life. Through memory (including computer memory), creativity, and foresight, the control room pools the assets of its internal studio workings, adjusts them, and sends them out through the head of its electronic nervous system, the transmitter.

Control room facilities exist to coordinate effectively the numerous video and audio elements that combine to produce a television program. Studio or remote camera shots must be monitored continuously and the desired camera shots switched to the on-air channel. Studio and intercom audio must be routed, and under continuous volume control. Basic equipment includes monitors for two kinds of signals, video and audio. These may be used for previewing and critiquing. Switchers select which studio equipment will be fed to the air. Special effects systems serve both audio and video. Cartridge (cart) playback units and recorders are found in the main control room or in an adjoining subsidiary control room.

Some control rooms are located adjacent to the studio, separated from the studio by a large glass window so that action and hand signals can be seen readily. Other control rooms are located in another part of the building, connected to the studio only by camera monitors, audio monitors (loudspeakers), and an intercom.

The studio control room has three basic areas: video, audio, and program. Each function should have enough operating space. The director must be aware of everything going on related to the program. He or she must see all the shots of the cameras and must be able to hear what is being said on the mike. Most importantly, the director must be in contact with the camera operators to give instructions for shots. The program control board consists of a monitor for all cameras, a preview monitor, and a program outmonitor. There are intercoms to all cameras and a monitor for the film chain/videotape. Seated next to the director is the *technical director* (TD).

CONTROL EQUIPMENT: VIDEO

Monitors

Each studio camera, film pickup chain, and remote camera is represented on a control room picture monitor that looks like a home television receiver without control knobs. There are also preview or preset monitors and on-air or program monitors. The preview monitors allow control room directors to preview and perfect a camera shot before it goes on the air and is seen on the program monitor. Because of the great number of monitors needed for all the video sources, monitors are only about 12 inches wide. Usually, only the preview and program monitors are color; color is more expensive and it is easier to determine picture quality in black and white.

Many studios have additional, separate monitors for engineering control, with one monitor for each camera. These monitors determine that the color, scanning, light, and voltage levels are correct. These monitors permit production and engineering personnel to carry on their functions without disturbing each other, which might happen if they shared the same monitors.

Control Room Switchers

Control Production Switcher

The switcher is the basic tool of the TD. Computerization and digital technology provide the capability for extensive visual creativity. The control room switcher consists of rows of buttons that allow the TD, upon cue from the director, to air any of the individual cameras and to produce special effects. There are individual buttons for each camera, separate buttons for each special effect, and a handle to control fading in and out (see Figure 5–12).

Most stations use complex switchers, some of which are custom built to their specifications. With automation, the number of inputs is unlimited and controlled only by the size of the switcher. Video and audio can be called up as well by master control from sources not in the studio, and include video carts, slides, and film from telecine.

In smaller stations some of the master control work may take place in the studio control room. For example, one program may be aired while another is being taped. As stations increasingly convert to digital switchers, control rooms and their equipment are getting smaller.

Routing Switcher

Another type of switcher is the routing switcher, which feeds and switches audio and video signals from any camera or switcher to those in other control rooms. It is an essential ingredient of broadcasting, supporting the master control switcher.

Production Switchers

Component video spawned production switchers that can handle the increasingly used analog component signals. Component

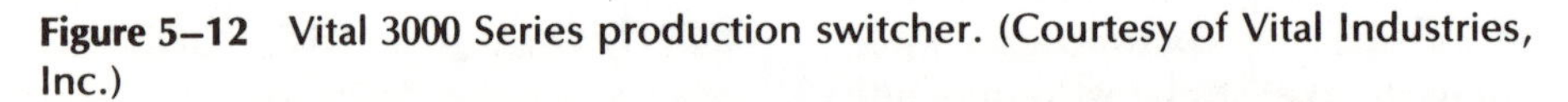
Figure 5–12 Vital 3000 Series production switcher. (Courtesy of Vital Industries, Inc.)

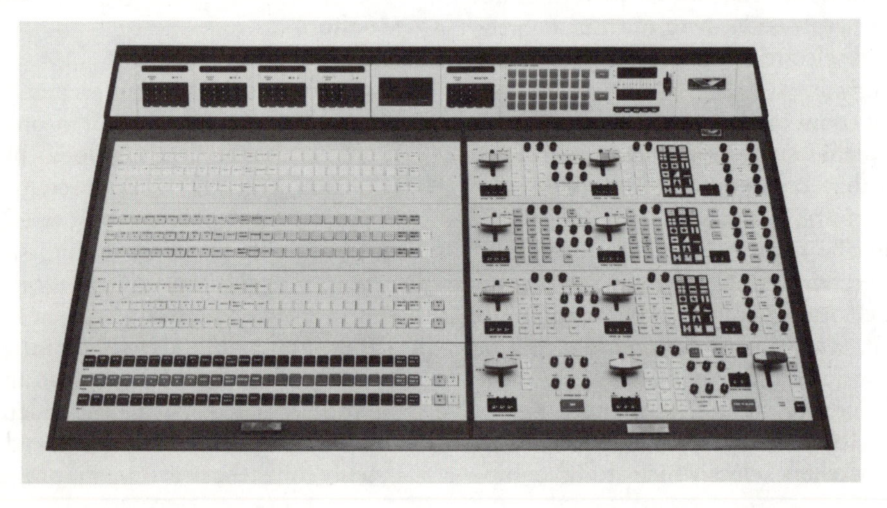

switchers are being developed that will work with any component format (see Figure 5–10). Another type, encoded production switchers, are used especially for encoded chroma keying.

LIGHTING

Adequate light in the television studio is vital to successful production. Complete studio lighting packages include a memory control console, computerized dimming systems, and lighting switchers.

Lamps used for production lighting range in power from 100 watt *baby inkies* to 10,000 watt *brute arcs*. The *tungsten-halogen* or *quartz* lamp, smaller and lighter in weight than previous lamps using comparable wattage, is predominantly used in production. It has a longer life and produces more efficient light, which is necessary for color, than do other types. It also maintains a constant color temperature or light intensity for the life of the lamp, and there are no light-intensity color changes on the recorded program—a particular concern when editing tapes.

The high-intensity HMI lamp was introduced in the 1970s. HMI lamps are filled with mercury, argon, and various rare earth elements to provide light of an intensity and a color temperature that approaches natural daylight. They produce several times as much light as do the tungsten filament lamps of the same wattage. Good for outside use, HMIs are, however, very large and require transformers that create a bulky package.

Portable lighting is essential today because so much production occurs outside of the studio. Field generators range from 500 watt portable units up to 900 watt mounted power plants. Field projects may also use mobile lighting trucks, such as Muscos, which can lift their light pods of sixteen 6,000 watt lamps as high as 150 feet. Electronic news gathering requires portable units because shooting does not always occur under satisfactory light conditions, and large lighting support may not always be available. Power for ENG lighting is provided by lightweight battery belts, some equipped with a tungsten-halogen source. Accessories for lighting systems include gel frames, barndoors, motion picture adapter cables, light stand mounts, and battery packs and chargers.

One must keep in mind the integral relationship between lighting and the other production elements: the need for the camera to produce a picture of transmission quality; the avoidance of shadows by sound equipment; the distortion sometimes suffered by microphones in the direct heat of strong lamps; and the high-pitched sounds that radiate when electric dimmers create movement of the lamp filaments. Video and sound and lighting are integrally related for good production.

AUDIO

Television's audio quality reception has not yet reached the standard of video reception. Although television's audio is distributed by FM signals, which are superior to AM signals, the public's ear has become attuned to the superb fidelity produced by advanced recording methods and new audio technologies. Not only is "sweetened" audio for television available in stereo and quad, but digital audio used for the compact disc has great potential; however, until stereo television receivers are manufactured with improved loudspeaker systems, the audio quality cannot be accurately reproduced. For example, digital broadcasting requires additional bandwidth while most television sets are still not designed to receive even stereo, much less digital audio, even though many television stations are broadcasting either a stereo or a synthesized stereo signal.

Just as is television's visual signal, the audio of television is radio transmission whereby sound waves are converted into electrical signals that modify a carrier wave by which they are transported through the air to home television receivers. There are many challenges in audio production. Sound is often reflected, bouncing off in all directions. There are problems of desirable or undesirable high-frequency to low-frequency balance and reception. Not only must an audio operator and director understand acoustics, but also what types of microphones emphasize high or low frequencies, and what determines ideal microphone placement.

Microphones

The microphone is the first electronic link in the communication chain between source and receiver. Basically, the microphone converts sound waves to electric current, using either velocity or pressure.

Dynamic Microphones

The *dynamic* microphone, the most rugged and frequently used, is a pressure mike. A *pressure mike* converts changes in air pressure to electric signals and consists essentially of a diaphragm enclosing a cavity that is sealed off from the air. Dynamic microphones are constructed with one or a combination of all pickup patterns. The *shotgun* microphone, used to pick up sound from distant sources, is a class of dynamic microphone with a supercardioid pattern and is very directional. The dynamic tends to screen out some of the muddy low frequencies of bass sounds, thus emphasizing the higher frequencies.

The pressure-gradient electrostatic microphone is another type of dynamic, referred to as a *condenser* microphone. The condenser contains a tiny piece of mylar film stretched over a solid backplate with an air space separating the two, thus forming a variable capacitor inside the microphone. (The word *condenser* is the former name for capacitor.) The mylar diaphragm moves toward and away from the backplate, thus changing the capacitance and creating the electrical signal, which is then amplified for distribution. Condenser microphones are manufactured in all of the existing directional characteristics. They have a wide and smooth high-frequency response and a low noise level. *Lavalier* microphones are a type of condenser.

Velocity Microphones

Velocity microphones operate on the speed (velocity) of the sound entering the microphone. The principle component of the velocity microphone is a corrugated aluminum ribbon suspended between the poles of a powerful horseshoe magnet. The speed with which the sound enters the microphone makes the ribbon vibrate within its magnetic field to reproduce the original sound. These mikes are often referred to as *ribbon mikes*.

Extremely sensitive, the velocity microphone has a warm, mellow quality and thus is often used for music pickups. Generally, it is found only in the studio because the delicate internal ribbon is extremely sensitive to shock or sudden gusts of air. Velocity microphones can be either bidirectional or unidirectional. This type of microphone tends to emphasize low frequencies and to screen out high frequencies.

Wireless Microphones

Wireless microphones are designed to be used in situations where the presence of microphone cables or lack of electricity would impede production. They operate through a built-in FM transmitter that can be picked up a short distance away by a receiving tuner set to pick up the frequency of only that particular microphone. Wireless microphones are also used when a studio longshot is needed and the cables present a visual problem. The performer wears the transmitter concealed in a pocket or underneath clothing, linked to the microphone by a lightweight wire. One type of transmitter is so small that it even can be concealed inside the human ear. The microphone and transmitter may also be combined into a single unit. The place where the wireless microphone is used must be checked out in advance to determine possible acoustical interference and to assure quality reception.

Lavalier Microphones

The arrival of television dictated development of smaller microphones for situations where the presence of a large microphone would be distracting. Sony was the first to introduce the *lavalier* microphone, particularly for broadcast news and interview use (see Figure 5–13). Lavaliers are designed to be resistant to external noise and ambient conditions. In addition, the cables to which the microphones are attached have been strengthened to resist breaking caused by overflexing. They can be powered by regular AA size batteries. One Sony model is the size of a fingernail, with a diameter of 5.6 mm. The tiny mike offers a complete frequency response rate, ranging from 40 to 20,000 Hz. Accessories for lavaliers include microphone clips, microphone pins, and urethane and metal wind screens.

Microphone Patterns

There are four basic pattern configurations to microphones and numerous variations and combinations. *Unidirectional* has a narrow cone of acceptance and is used for situations where a tight pickup of sound is needed, blocking out extraneous sounds and noise. *Bidirectional* provides live pickup on two opposite sides and is ideal for across-the-table interviews. *Omnidirectional* has a 360-degree pickup and is used where a large range of sounds are sought, such as with news, sports, large audiences, and other remotes. *Cardioid* has a heart-shaped pickup pattern.

A ratio of 4:1 is the ideal pickup discrimination between the origination signal (incident sound) and unwanted noise (ambient sound). The farther the microphone is from the source of the sound, the more careful must be the selection of the correct type of microphone to maintain the desired sound ratio.

Microphone Mountings

Microphones can be mounted on a variety of stands, hung around the neck, or concealed in clothing. A microphone must be firmly screwed to a stand if it is not to be hand-held. For combination stand and hand-held use the mike is mounted on a stand in a special holder from which it is easy to remove when necessary. For studio work the boom on wheels used to be the principal means to hold microphones suspended above the performers or the action. Wireless micro-

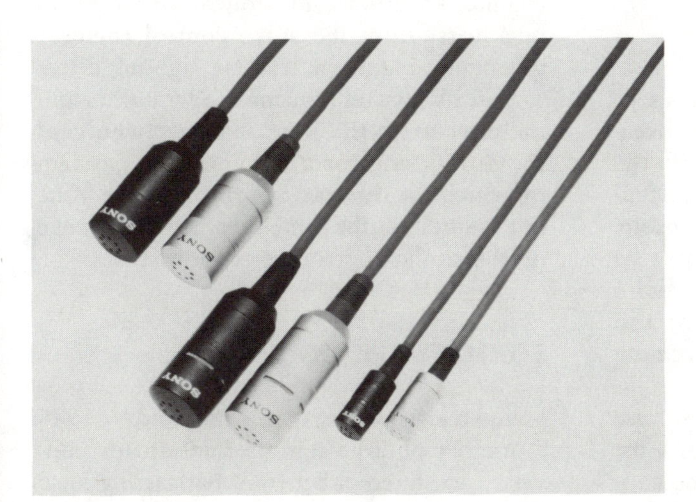

Figure 5–13 Sony ECM-55, ECM-66, and ECM-77 lavalier microphones. (Courtesy of the Sony Corporation of America.)

phones are gradually replacing the bulky boom. When a boom or hung mike is needed, to pick up audience response, for example, the cardioid is preferred because it is narrow enough to discriminate between incident and ambient sound but is open enough to pick up more than one source. More sensitive microphones require that cameras are as quiet as possible. Noises that a microphone might pick up include the sounds from a camera's zoom mechanism or its cooling fan.

AUDIO CONTROL EQUIPMENT

The console operator who handles audio for television must gather audio from all inputs, then control, process, and combine them so that they support but do not dominate the video. Inputs include studio microphones, control room microphones, taped audio music or sound effects, telecine, and remote sources.

Audio for television must be loud enough for its purpose, whether foreground or background, but never overmodulated to the point of distortion. Proper balance must be maintained between sound to sound, sound to voice, or sound to music, according to the creative director's wishes. Separately produced audio items must be matched so that the ear seems to hear all at the same level.

Audio Operations

When possible, the audio control area is separate from but adjacent to the video area. Whether apart or in the same room with the video, the audio console, carts, and related equipment have their own desk and console. In addition to controlling audio from the studio, the audio engineer controls audio from the control room to the studio, and prerecorded material and audio from remote locations.

Audio consoles usually are digital and use *slide faders* to control volume, although some engineers believe that boards using the conventional round *pots* are easier to maintain. An increasing number of consoles are able to handle stereo. Volume control and switching are the primary functions of the audio operator, who (1) presets levels of audio by auditioning sources of live sound to check whether microphones are working, (2) restricts unwanted sound, and (3) balances sound. The main fader should not be touched once it is set with presumably the best quality for all other volume controls.

Several kinds of consoles provide audio to accompany video, some for in-house production, some for direct broadcast, some to combine broadcast and production, and some to mix sounds for production. Consoles not only handle stereo and multitrack production, but are expandable to handle from sixteen to eighteen inputs. Intelligent digital faders and several levels of automation enable radio consoles to interface with video editing systems and television facilities. Microprocess-controlled switchers handle the input and output of stereo routing and mixing.

Intercom

The intercommunication system is essential for personnel to coordinate their efforts: circuits from the director and technical director to the camera operators and video control technicians; from audio control to the boom operators; from the video control engineer to camera operators, telecine, lighting director, and VTR and film chain operators; from the director to the floor manager who cues the action; and for a public address system by which the director can speak to everyone in the studio at the same time. All these make up the studio/control intercom system.

STEREO TELEVISION

Stereo television began in 1984, and by 1986 a number of large stations had already converted to stereo either by broadcasting local productions in true stereo, or synthesizing monaural audio to sound like stereo when programs originated from monaural sources.

True stereo involves two completely separate channels that are aired and received simultaneously. When this is not possible, the signal is *multiplexed;* that is, the original signals from two different mikes or sources are encoded in a multiplexer and sent to the broadcast antenna as a signal, and a decoder pulls apart the combined two signals before it reaches the receiver so that the originating sounds are heard through the right and left loudspeakers.

Multichannel Television Sound

Development of *multichannel television sound* (MTS) is the key to true stereo broadcasting. Merlin Engineering introduced a system that put stereo audio into the vertical interval of the video signal. MTS can be used on any straight cable run, as well as for microwave or satellite links.

Miking for Television Stereo

Stereo adds spice and dimension to the visuals, but requires a director to plan camera shots with stereo audio in mind, and necessitates longer postproduction time.

Some sitcoms shoot in limited stereo; that is, not with stereo dialogue, but by splitting the audience audio into left and right sound tracks and later adding stereo music and stereo sound effects. While the laugh track in monaural is placed very close to the punchlines, in stereo it is set further back to create a feeling of space. A typical sitcom show with a live audience may use 12 microphones for the audience, separated for stereo. The videotape is recorded with the set's audio on the top track and the audience sound on the bottom track. Sound effects done on the set, such as doorbells or phones, are placed on the audience track. Other effects are mixed in during postproduction.

The audio mixer for a live stereo television show must be able to see where the talent is, and to hear the sound through a loudspeaker rather than through less acoustically accurate headphones.

Studio Acoustics for Stereo

If stereo audio is faulty, then dimension, color, and texture of the sound will be wrong. Studio acoustics are more critical for stereo than for monaural audio. One way to fight acoustical problems is to put absorbent material on one of the two parallel walls in order to knock down standing-wave echoes in the studio's camera areas. A live-end dead-end (LE-DE) type room is best for live stereo television production and is basically anechoic (without echo), thus minimizing or eliminating reflections from the surroundings that can feed into the audio and mix with the recorded sound. Precise measurements can be made with *time delay spectrometry* (TDS). The goal is to create a *Haas zone,* with the initial reflection from the direct sound off the rear wall of the acoustic room made inaudible, and to have any energy remaining after those reflections to decay evenly over time.

Haas Effect

The goal of the Haas effect is to shift stereo images without unbalancing levels. A Haas zone is based on the Haas effect, which occurs when the human brain masks echoes and delays of sound that arrive approximately ten to twenty milliseconds after the original or direct sound, fooling the ear into thinking that the first of two closely time-spaced signals is dominant. To illustrate: if a musical instrument is recorded at equal levels in each channel, with one channel delayed a few milliseconds, the listener will perceive the reproduced sound as coming from the channel that offers the leading signal. The advantage of this effect is that you can place an instrument or voice on one side of the stereo stage without having to lower the level in the opposite channel. A room using these principles of acoustics will not have any reflections or reverberations feeding into and coloring the final audio product. What is heard in such a room will be accurate to the actual sound.

Stereo Television Programming

By the late 1980s many major prerecorded programs were being produced in stereo, partly to improve the presence of the audio and partly for ready identification as stereo when the shows go into syndication and stereo television sets become commonplace. Stereo appears more in music than in dialogue. "The Tonight Show" in 1986, for example, still used the table microphone and overhead boom to pick up all the interviews and dialogue in monaural audio, while the miking of the audience, bands, and singers was in stereo.

Digital technology has facilitated the use of stereo in television. One of the first sitcoms to be produced and broadcast in stereo was "The Cosby Show." Acousticizing the Cosby set produced design techniques for the entire recording industry. NBC even renovated an audio postproduction room especially for the Cosby program.

Secondary Audio Program Channel

In addition to the obvious advantage of stereo to add reality to a scene, stereo also brings with it the potential for multichannel programming by means of the *secondary audio program channel* (SAP). SAP made early progress with bilingual audio, particularly on syndicated programs. For example, English can be heard on one channel while a foreign language is heard on the other. In communities with large Hispanic populations SAP has been used in newscasts as well as in situation comedies. Specially trained audio technicians and translators are needed for SAP production. The translator prepares scripts in the foreign language and the technician edits scene by scene in order to match audio to video so that the foreign language version is not out of sync with the video. To broadcast SAP a television station must be equipped for stereo, and then add an SAP generator and TV stereo exciters. The audio is distributed either on the television station's subcarrier or its aural carrier, depending upon the system in use.

REMOTES AND SATELLITES

Satellites are important in the distribution and reception of television and audio signals. Remote ENG production/transmission, for example, depends upon terrestrial microwave and satellites for the most efficient news operations.

Mobile Television Communications

A typical mobile unit handles satellite news gathering (SNG) as well as ENG with onboard power, editing and microwave communication capabilities, and a deployable antenna. Advanced mobile units have digital carriers that permit encrypted communications between a remote or base station, in addition to the ability to access any satellite, transponder, or frequency.

Mobile Production Facilities

Mobile production units range in size from small 1 ton vans to 45 foot trailers and support not only news gathering but production field work, commercial shoots, sporting events, parades, and other large outdoor happenings. Remote trucks are designed to serve as master and operating control rooms on wheels (see Figure 5–14). Electronic news gathering vans are used for limited production in addition to news gathering. The smaller units are owned by individual stations. Large mobile units may not be used enough to justify ownership and are frequently leased as needed. Large mobile units

Figure 5–14 A mobile production trailer, forward view. Specifically, this is the instant replay tape area of an ABC Sports tape trailer. (Courtesy of the Centro Corporation.)

may contain multiple cameras—at least three hard cameras and two minicams—and are self-contained with 1 inch tape equipment for production purposes. Smaller trucks use 3/4 inch, with many switching in the late 1980s to the Beta 1/2 inch format.

Helicopter

The helicopter is a useful mode of news gathering when terrain or the type of event prohibits use of a mobile unit. The helicopter teams with a mobile unit to relay the story back to the newsroom via the small sending antenna on the truck.

Terrestrial Microwave Transmission

Terrestrial microwave transmission is used routinely to relay back to the station live regional news, sports, and entertainment feeds. Microwave also serves stations as the first/last mile link to and from satellite uplinks and downlinks, facilitating live program feeds from anywhere in the country or the world. Complementing fixed microwave facilities throughout the country are mobile microwave transmission vehicles. Some carry only the main necessities for signal transmission; others are custom designed for remote broadcast production, complete with everything from ENG cameras to stereo audio recording equipment. In hilly terrain and in areas with many tall buildings, where it is impossible to operate transportable C-band uplinks directly, terrestrial microwave units can link an event to transportable uplinks located miles away. For example, Gannett Broadcasting covered the 1984 Olympics, with its multiple settings and hilly terrain, by equipping a hotel room with videotape playback and editing facilities, adding a portable terrestrial microwave unit parked outside the Los Angeles Coliseum, and feeding live coverage and videotapes via microwave to a VideoStar Connections transportable uplink at the Burbank airport, which in turn fed the signal to the satellite.

Satellite News Gathering

Satellite news gathering extends the radius of a station's news department far beyond its terrestrial microwave ranges. It permits a news gathering network to be set up among regional and nationwide stations to feed regional stories to the network. To support SNG, audio and video signals are fed to a satellite transponder by an uplink earth station using either C-band or Ku-band frequencies. (The C-band uses the same frequency spectrum, 3.7–4.2 gigihertz, as most terrestrial operations, and requires a large satellite receiving dish of at least 35′ in diameter, while the Ku-band is in the less crowded 12.20–12.70 gigihertz spectrum and can use a dish as small as 5′ in diameter.) The transponder puts the signal on an appropriate frequency and sends it back to the proper satellite earth terminal by a downlink *television-receive-only receiver* (TVRO), which then sends the signal to the station by terrestrial microwave.

The C-band operates in the megahertz band and the Ku-band is located in the gigahertz group (megahertz means one million cycles per second; gigahertz means one billion cycles per second). Terrestrial microwave also broadcasts from the gigahertz band. The more cycles per second, the smaller the wavelength and the smaller the required sending and receiving antennae. C-band equipment is relatively inexpensive and has sturdy technology, but requires careful coordination of both the uplink and downlink sides with the terrestrial microwave facilities which may share the same frequency. Ku-band is more costly to operate and the equipment is more sensitive (it can even lose its signal when it rains), but it does not suffer the coordination problem of C-band. Ku-band uses satellites with high power and higher gain, thus allowing good quality signals to be sent and received with smaller antennae both on the ground and on the satellite. In the late 1980s uplink antennae were getting smaller, satellites more numerous, and more stations were choosing the Ku-band.

The quality of a signal is related to the size of the antenna (see Figure 5–15 for examples of satellite antennae). A small antenna, while less susceptible to interference because it is easier to shield, may also receive less of the actual signal from the satellite and, consequently, may pick up more noise. A motorized polar mount enables an antenna to change position and switch channels at the touch of a button. A programmable position controller can store up to sixteen satellite positions and can be remotely controlled by either telephone or a station-to-transmitter microwave link (STL).

Satellite News Gathering Production Audio Channels

In addition to the video and audio sent back to the station for broadcast, there are a number of audio channels not intended to be heard by the public but that are essential for microwave and SNG to work. One channel is needed for production coordination, a second for the homebase control room to pick up the camera and sound, a third for the program producer to talk to the field reporter, and other channels for additional op-

Figure 5–15 Three different sizes and types of satellite antennae. Microdyne 5 meter antenna and motorized polar mount *(left),* 3 meter antenna *(center),* and CH-14 conical horn antenna *(right).* (Courtesy of the Microdyne Corporation.)

erational needs. Communication must be available during travel, set-up, and transmit periods. Although only one circuit is needed during travel time, the set-up period may require five or six channels, which then must remain open during transmit time (Figure 5–16 illustrates SNG communication circuits options).

Flyaways

The higher the frequency the smaller is the required sending/receiving antenna. In 1986 the smallest antenna was 1.8 meters in diameter. Smaller antennae have made the flyaway possible. A *flyaway* is a portable uplink making use of modular construction that allows the antenna to be assembled onsite out of airline baggage-checkable cases. It enables one person to handle a distant story single-handedly. The Rover flyaway, for example, breaks into four modules: one antenna and three electronic cases, each weighing 130 pounds.

AUTOMATED SYSTEMS

Automation, initially of special use in post-production, moved rapidly into all other aspects of station operation. By the late 1980s television was making a transition from complete analog to complete digital systems, with many stations having automated systems all the way through master control.

Engineering and Technical

Master Control

Master control is the heart of a station's operation. It links everything together. It houses the master control switcher which works together with the distribution switcher and machine control switcher. In master control, or linked to it electronically, are electronics for studio and telecine cameras, character generators, graphics, still-store, switchers, and intercom systems.

An example of the scope of automated operations possible in master control is Alamar Electronics' computer-driven automated switcher system, which can do all of the following: provide network delay programming; print a station log as events are aired; automatically cue commercials, public service announcements, promotional spots and station IDs; control more than thirty VTRs and telecine interfaces; provide random ac-

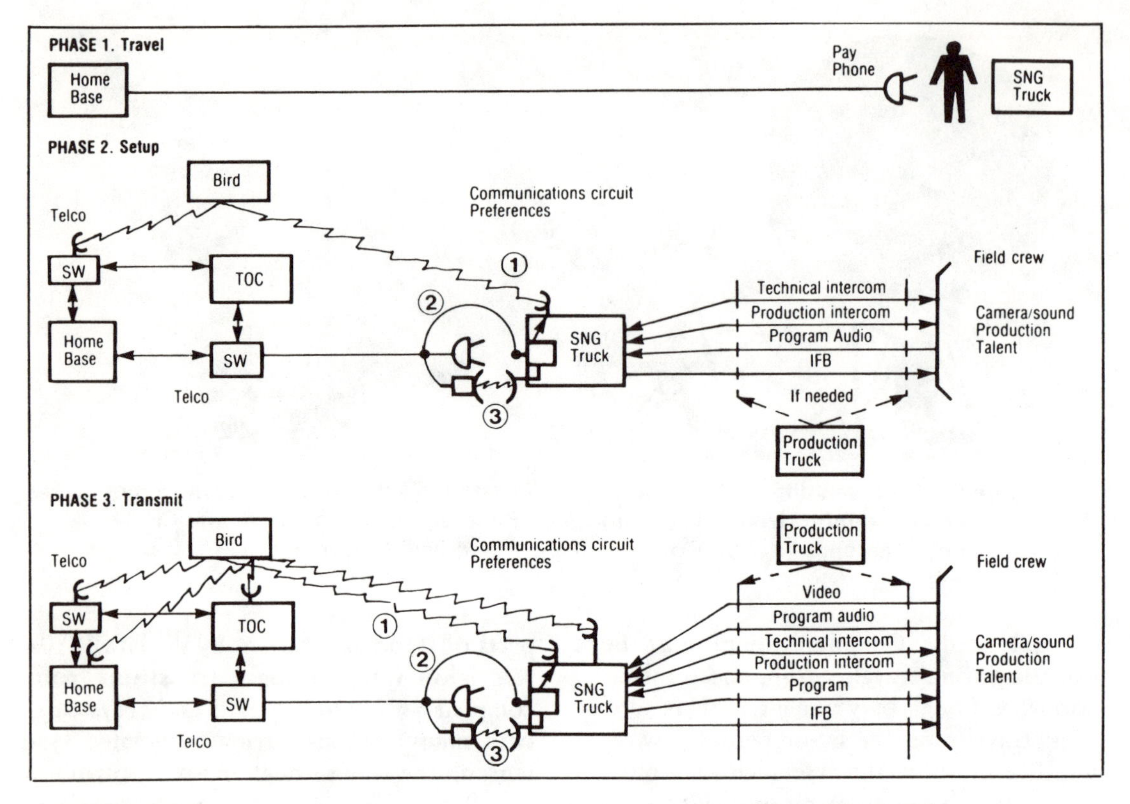

Figure 5–16 SNG communications circuits options. (Courtesy of the Sony Corporation of America.)

cess to multiple spots on tape; handle on-line event editing; and provide on-screen operator prompting. It is even equipped with a *help* key to aid any level of user.

Remote Control

A remote control microprocessor can monitor as many as ninety-nine remote sites with multiple control terminal capability so that an operator can delegate control from master to master as needed. Projectors and telecine transports can also be remotely controlled from master control.

Production and Postproduction

The multiple event record and playback system (MERPS), using the M-II tape format, was a new development in the late 1980s. It operates in both VHS and Beta formats and has been favorably received. Video cart equipment has moved down from 2 inch quad to varied widths as narrow as 19 mm. Another automated editing tool, the ACAR-225, provides one channel of digital composite video plus four channels of digital audio, all on a 19 mm cassette. As many as 256 cassettes can be on-line at any time while a library database can identify more than ten thousand cassettes.

Programming and Logging

The computer used in an automatic broadcasting system is a specially developed real-time machine that can store large amounts of information inexpensively. Depending on the size of the memory, a few days, a week, or up to a month's worth of programming can be held in the computer at one time. In addition, once a number of commercial slots have been programmed into the day's schedule, the computer can be asked for information on what spots are unfilled. This information can be requested from a distance using normal telephone circuits.

Computers also note the time and transmission of information and print up station logs. Program logs are sometimes determined up to thirty days in advance. The computer-prepared log may include the time the item goes over the air, the duration, whether it is a video or audio source, the number of the film or tape, a description of the item, and whether it is a commercial. The computer may later print up another log showing what actually went on the air. If items to be aired are changed at the last minute, they must be recalled from the computer and others inserted. When an item that is due to air is not found by the computer, the technical director must insert something else. The technical director can manually override the system at any time.

HIGH-DEFINITION TELEVISION

Another new development in the 1980s expected to have profound impact on television is HDTV. Pioneered by NKH (Japan Broadcasting Corporation) with Sony, Panasonic, and Ikegami, HDTV uses 1,125 horizontal lines to scan the picture as compared to the 525 lines of the NTSC system, thus providing a much better picture, comparable to 35 mm theatrical film.

HDTV has been applied to closed systems, videocassettes, and videodiscs. It has also been of great value in medical applications. Film producers agree that HDTV would provide "electronic films" of high enough quality for theatrical showing.

A major hurdle is the 30 MHz bandwidth required for HDTV transmission, while the bandwidth authorized for broadcast VHF and UHF channels is only 6 MHz. One solution experimented with in the late 1980s is to compress the signal with digital component standards. Another solution (see Figure 5–17) involves alternate ways in which programs might be distributed to homes. One method is by *direct broadcast satellite (DBS)* transmission direct to the home base. Another is by cable, using optical fiber links rather than present coaxial systems, which might not be able to transmit wide-band television without serious signal degradation. An additional method would be through local stations, which would not actually broadcast HDTV, but would convert HDTV tapes to compressed NTSC for broadcast purposes. *Multipoint distribution systems* (MDS), which uses microwave, is another alternative.

In 1985 a national U.S. standard was agreed on when the HDTV Technology Group of the U.S. Advanced Television Systems Committee (ATSC) recommended a 1,125-line, 60 Hz, 2:1 interlace standard with an aspect ratio of 5:333:3 for studio production of HDTV programming. This is essentially the same standard originally developed by NKH. In 1986 high-frequency switching systems were developed for HDTV. It is expected that HDTV will be introduced in theatrical films, videocassettes, and videodiscs. With the resolution of compatibility problems, HDTV will likely reach the public through broadcast television in the 1990s.

TECHNICAL OVERVIEW

Television transmits its broadcast signal using AM radio waves for video and FM waves for audio. The visual portion is transmitted by a scanning process that broadcasts thirty 525-line frames every second. Each of the 525 horizontal lines varies in brightness according to the image it is transmitting; the changing brightness produces a varying voltage that is transmitted to the home receiver, which reconverts the changing light levels back to video.

Color television is a combination of properties of light that control brightness, hue,

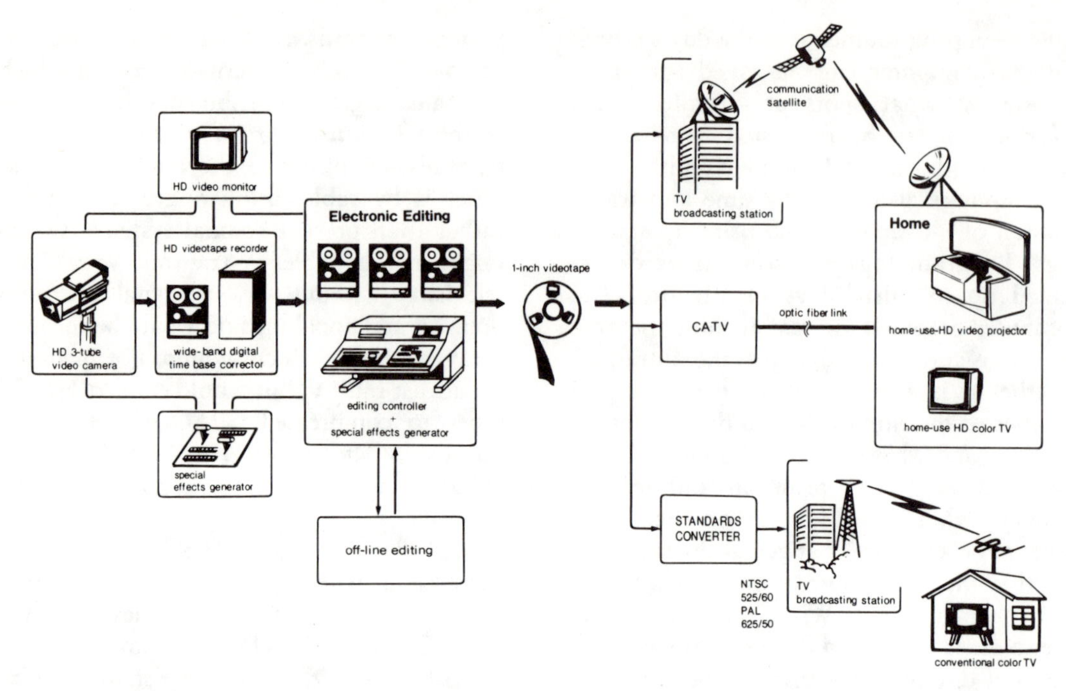

Figure 5–17 High-definition video system future broadcasting applications. (Courtesy of the Sony Corporation of America.)

and saturation, using the primary colors of red, green, and blue, which can be combined to create any color. Brightness is the only portion of a color signal that is picked up by a black and white receiver.

Electronic television cameras rely upon their pickup tubes to convert the optical image into an electronic signal capable of being broadcast. Some cameras need only one tube, others require more. Pickup tubes vary in size, as well as in the makeup of the tube's photoconductor. Some electronic cameras are designed exclusively for studio use, others for field production or news; some can handle both studio and field. While tubes continue to improve, the CCD, developed in the 1980s, appears to provide qualities superior to tubes in certain aspects.

Professional television cameras use three tubes to capture the red, green, and blue components of a picture. The higher the resolution, the better the picture. Resolution is expressed in the number of horizontal lines needed to make up an image. Development of HDTV has led to the manufacture of some studio cameras capable of resolving 1,125 scan lines per picture. As more computer elements and chips are added, cameras continue to become smaller, lighter, more automatic, and more efficient in energy saving. Electronic news gathering cameras have generally replaced 16 mm cameras for television news location shoots. The recording camera, a self-contained portable video camera, is used primarily for ENG. The advantage of the camcorder is that the video can be played back immediately and viewed in the camera's viewfinder.

There is a wide variety of video lenses for different kinds of shoots, ranging from the all-purpose lens for smaller video cameras to large lenses for specialized needs. The way a camera is used determines the type of lens chosen. As cameras become more sophisticated, so do lenses, many with automatic features, using miniature computer systems built into all parts of the camera processing unit. The solid state sensor is replacing the pickup tube in many instances.

Videotape used in video recorders, cartridges, and cassettes continues to become more sensitive and requires special care to maintain top recording quality. Videotape recorders are usually identified by the width of the tape they use. Digital signals offer more fidelity and less distortion than do analog systems. Digital technology is used in aural and visual applications as well as in special effects. Coming into the market are laser disk recording systems and hard computer disk systems. Editing of videotape is performed through use of time base correctors using either assemble editing or insert editing, on-line or off-line.

Film and slides are aired through telecine. Modern telecine equipment uses digital processing and CCD image sensors to produce high-quality video pictures.

There are two types of control rooms: master control and studio control. Master control coordinates every source for air; the studio control room is used to direct the shooting of a particular program or spot for live or prerecorded purposes.

Adequate light is essential to successful studio production. Selection of appropriate lamps, use of memory control consoles, computerized dimming systems, and lighting switchers are all part of the lighting systems found in studios.

Microphone selection depends upon type of voice, music, sound, source, distance, and desired effect. An operator must know the different pattern configurations of each type and how to achieve incident sound while avoiding ambient sound. Acoustical control, always essential in television studios, is critical in stereo television production. Until speakers in home receivers are upgraded, the public will not hear fidelity stereo.

Satellites are used increasingly to distribute and receive television and audio signals, for ENG and for special events. Sometimes mobile units handle SNG in addition to ENG, and the largest mobile production units can serve as master and operating control rooms on wheels. Terrestrial microwave transmission is used regularly to relay live regional news to the station, to provide sports and entertainment feeds, and to serve as links to and from satellites. Satellite news gathering extends a station's news department far beyond its terrestrial microwave range, using either C-band or Ku-band.

Automation continues to grow in all aspects of equipment and control rooms, to serve production, engineering, and daily needs of programming and logging.

BIBLIOGRAPHY

Alten, Stanley R. *Audio in Media.* Belmont, CA: Wadsworth Publishing, 1981.

Bermingham, Alan, et al. *The Small Television Studio.* Boston: Focal Press, 1976.

Browne, Stephen E. *Videotape Editing: A Postproduction Primer.* Boston: Focal Press, 1989.

Carlson, Sylvia and Verne Carlson. *Professional Cameraman's Handbook.* 3d ed. Boston: Focal Press, 1980.

Carlson, Sylvia and Verne Carlson. *Professional Lighting Handbook.* Boston: Focal Press, 1985.

Fielding, Raymond. *The Technique of Special Effects Cinematography.* 4th ed. Boston: Focal Press, 1985.

Hilliard, Robert L., ed. *Radio Broadcasting.* 3d ed. New York: Longman, 1985.

Millerson, Gerald. *The Technique of Television Production.* 11th ed. Boston: Focal Press, 1985.

Nisbett, Alec. *The Technique of the Sound Studio.* 4th ed. Boston: Focal Press, 1979.

Oringel, Robert S. *Audio Control Handbook.* 6th ed. Boston: Focal Press, 1989.

Robinson, Joseph F. and Stephen Lowe. *Videotape Recording.* 3d ed. Boston: Focal Press, 1981.

Watkinson, John. *The Art of Digital Audio*. Boston: Focal Press, 1988.

Wurtzel, Alan. *Television Production*. New York: McGraw-Hill, 1979

Zettle, Herbert. *Television Production Handbook*. 4th ed. Belmont, CA: Wadsworth Publishing, 1984

Periodicals with a special emphasis on the technical aspects of broadcasting include:

Broadcast Engineering, *Television/Radio Age*, and *Broadcast Management/Engineering*.

6

Writing

Robert L. Hilliard

The television writer aims at an audience that at one and the same time is very small and very large, that has much in common and almost nothing in common, that is a tightly knit group and a disunified mass.

Millions of people may be seeing the material developed by the writer. Yet, any one group within this vast audience is apt to be a small one—usually a family group at home in everyday surroundings. The distractions of everyday life are constantly at hand, continuously operative, and likely to pull the individual viewer away from the program. The writer must capture the imaginations and interests of the audience as soon as possible. Each word and each picture must be purposeful, must gain attention and hold interest. Ideally, there should be no irrelevancies in the writing and no extraneous moments.

To make any single piece of material effective, the writer often tries to find a common denominator that will reach and hold as many as possible of the groups and individuals watching the more than 180 million television sets in use in this country.

Unfortunately, there is an acceptance of the *lowest* common denominator and a reliance upon a quantitative measurement. The cultural contributions of our mass media have become, for the most part, comparatively mediocre in quality and repetitive in nature.

THE MASS AUDIENCE

Television uses many of the techniques of the theatre and of the film. With its use of mechanical and electronic devices, television has more flexibility than the theatre but, because of the limitation of sight as opposed to the unlimited imagination of hearing alone, is not as flexible as radio. Nevertheless, television can combine the sound and the audience orientation of radio, the live continuous performance of theatre, and the electronic techniques of the film. It is capable of fusing the best of all previous communications media.

At the same time, television also has specific limitations. It is greatly restricted in production by physical time and space. Timewise, the writer cannot develop a script as fully as might be desirable. Actual program length, after commercial and intro and outro credit time has been subtracted, runs about twenty-one minutes for a thirty minute program and about forty-two minutes for an hour program. This limitation is a particular hindrance in the writing of a dramatic program. Spacewise, the writer is hampered by the limitation of the camera view, the limitation of settings for live-type taped television (*live-type taped television* refers to the taped program that uses the continuous action, nonedited procedure of the live show;

it is done as if it were a live show) and the comparatively small viewing area of the television receiver. The writer must orient the script towards small groups on the screen at any one time and make extended use of the close-up shot in studio produced shows. These limitations prompt the intimate, subjective approach in dramatic writing, used successfully in recent years in such programs as "Hill Street Blues," "L.A. Law," "St. Elsewhere," and "The Cosby Show."

Television combines both subjectivity and objectivity in relation to the audience, fusing two areas that are usually thought of as being mutually exclusive. Through use of the camera and electronic devices, the writer and director frequently may give the audience's attentions and emotions a subjective orientation by directing them to specific stimuli. The television audience cannot choose, as does the theatre audience, from the totality of presentation upon a stage. The television audience can be directed to a specific stimulus that most effectively achieves the purpose of the specific moment in the script. Attention can be directed to subtle reaction as well as to obvious action.

The writer not only faces a problem with the quality level of the material, but faces concrete manifestations of this problem in the selection of specific subject matter. Television writing is affected greatly by censorship. Censorship falls into two major categories: material that is *censorable* and material that is *controversial*. Censorable material, as discussed here, is that which generally is considered not in good taste for the home television audience, although this same material might be perfectly acceptable in the legitimate theatre or in films.

The FCC on occasion acts as an arbiter of public taste. The Communications Act of 1934, as amended, authorizes fines or license suspension for "communications containing profane or obscene words, language, or meaning . . . ."

Censorship of controversial material is of concern to the writer. Controversial material refers to subject matter that in the broadest sense might disturb any viewer. Such material might relate to any area of public thinking, including political, social, economic, religious, and psychological areas. "When a story editor says, 'We can't use anything controversial,' and says it with a tone of conscious virtue, then there is danger," observed Erik Barnouw.

There is a great danger to freedom of expression and the democratic exchange of ideas in American television because many of the media executives fear controversy. On the grounds of service to the sponsor and on the basis of high ratings for noncontroversial but mediocre entertainment, controversy has been avoided in too many cases.

In the 1950s countless writers, performers, and other TV personnel were blacklisted because unproven allegations concerning their loyalty made them controversial. And only a few years ago the "Lou Grant" show was canceled because intolerance by certain pressure groups of program star Ed Asner's political beliefs made him controversial.

The media are powerful. Commercials do sell products and services. News and public affairs programs have become significant factors in creating as well as reporting news and in influencing much of our political and social policies and actions.

The writers who prepare continuity and background material for programs dealing with controversial issues and events have the satisfaction—and responsibility—of knowing that they are participating directly in changing society. There are not too many professions in which one can accomplish this on such a broad and grand scale.

Theoretically, the writer can help to fulfill the responsibility of the mass media to serve the best interests of the public as a whole, can raise and energize the cultural and educational standards of the people, and thus strengthen the country. Realistically, the most well intentioned writer is still under the control of the network and advertiser whose first loyalties usually are directed toward their own interests and not necessarily toward those of the public. Occasionally, these interests coincide. The writer who wishes to keep a job in the mass media is pressured to serve the interests of the station. It is hoped that conscience will also enable the writer to serve the needs of the viewers.

BASIC PRODUCTION ELEMENTS

The writer must learn what the camera can and cannot do, what sound or visual effects are possible in the control room, what terminology is used in furnishing directions, descriptions, and transitions, and what other technical and production aspects of the media are essential for effective writing. There are six major areas the writer should be aware of: the studio, the camera, the control room, special video effects, editing, and sound.

The TV Studio

Studios vary greatly in size and equipment. Network studios, where drama series and specials usually are produced, not only have all the technical advantages of a television studio, but also have the size and equipment of a movie sound stage. Some individual stations have excellent facilities, others are small and cramped. The writer should be aware of studio limitations before writing the script.

The Camera

Whether the show is being recorded by a film camera or by a television camera on videotape, the basic movements of the camera are the same. Even the terminology is the same. The principal difference is in the form of the script. The examples below detail camera movement. In some instances style may be different: short, individual takes for the film approach; longer action sequences and continuous filming for the live-type taped television approach.

In either case, the writer should consider the camera as a moving and adjustable proscenium through which the attention of the audience is directed just as the writer and director wish. Camera movement may change the position, angle, distance, and amount of subject matter seen. There are five major movements the writer must be aware of and be prepared to designate, when necessary, in the script.

Dolly In and Dolly Out/Zoom In and Zoom Out

The camera is on a dolly stand, which permits smooth forward or backward movement. This movement to or away from the subject permits a change of orientation to the subject while keeping the camera on air and retaining a continuity of action. The zoom lens accomplishes the same thing without moving the camera.

Tilt Up and Tilt Down

This consists of pointing the camera up or down, thus changing the view from the same position to a higher or lower part of the subject area. The tilt is also called panning up and panning down.

Pan Right and Pan Left

The camera moves right or left on its axis. This movement may be used to follow a character or some particular action, or to direct the audience to a particular subject.

Follow Right and Follow Left

This is also called the *travel* shot or the *truck* shot. It is used when the camera is set at a right angle to the subject and either moves with it, following alongside it or, as in the case of a stationary subject such as an advertising display, moves down the line of the display.

Boom Shot

Originally used for Hollywood filmmaking, the camera boom has become a part of standard television production practice.

Equipment, usually attached to the moving dolly, enables the camera to *boom* from its basic position in or out, up or down, at various angles—usually high up—to the subject.

Basic camera movements are written into

Figure 6–1 Standard two-column television format.

VIDEO	AUDIO
ESTABLISHING SHOT.	DETECTIVE BYRON (AT DESK, IN FRONT OF HIM, ON CHAIRS IN A ROW, ARE FOUR YOUNG MEN IN JEANS AND LEATHER JACKETS, WITH MOTORCYCLE HELMETS NEARBY.) All right. So a store was robbed. So all of you were seen in the store at the time of the robbery. So there was no one else in the store except the clerk. So none of you know anything about the robbery.
DOLLY IN FOR CLOSE-UP OF BYRON.	(GETTING ANGRY) You may be young punks but you're still punks, and you can stand trial whether you're seventeen or seventy. And if you're not going to cooperate now, I'll see that you get the stiffest sentence possible.
DOLLY OUT FOR LONG SHOT OF ENTIRE GROUP. CUT TO CU. PAN RIGHT ACROSS BOYS' FACES, FROM ONE TO THE OTHER, AS BYRON TALKS.	Now, I'm going to ask you again, each one of you. And this is your last chance. If you talk, only the guilty one will be charged with larceny. The others will have only a petty theft charge on them, and I'll see they get a suspended sentence. Otherwise, I'll send you all up for five to ten.
FOLLOW SHOT ALONG LINE OF CHAIRS IN FRONT OF BOYS, GETTING FACIAL REACTIONS OF EACH ONE AS THEY RESPOND.	(OFF CAMERA) Joey? JOEY (STARES STRAIGHT AHEAD, NOT ANSWERING.) BYRON (OFF CAMERA) Al? AL I got nothin' to say. BYRON (OFF CAMERA) Bill? BILL Me, too. I don't know nothin'. BYRON (OFF CAMERA) O.K., Johnny. It's up to you.

VIDEO	AUDIO
	JOHNNY
TILT DOWN TO JOHNNY'S BOOT AS HE REACHES FOR HANDLE OF KNIFE. TILT UP WITH HAND AS IT MOVES AWAY FROM THE BOOT, INTO AN INSIDE POCKET OF HIS JACKET. CUT TO MEDIUM SHOT ON BOOM CAMERA OF JOHNNY WITHDRAWING HAND FROM POCKET. BOOM INTO CLOSE-UP OF OBJECT IN JOHNNY'S HAND. (ORDINARILY, A BOOM SHOT WOULD NOT BE USED HERE. A ZOOM LENS WOULD BE EASIER TO USE AND AT LEAST AS EFFECTIVE.)	(THERE IS NO ANSWER. THEN JOHNNY SLOWLY SHAKES HIS HEAD. UNPERCEPTIBLY, BYRON NOT NOTICING, HE REACHES DOWN TO HIS MOTORCYCLE BOOT FOR THE HANDLE OF A KNIFE. SUDDENLY THE HAND STOPS AND MOVES UP TO THE INSIDE POCKET OF HIS JACKET. JOHNNY TAKES AN OBJECT FROM HIS POCKET, SLOWLY OPENS HIS HAND.)

the following scripts. In Figure 6–1, using the standard television format, the writer would not ordinarily include so many camera directions, but would leave their determination to the director. They are included here to indicate to the beginning writer a variety of camera and shot possibilities. The left-hand column, as shown here, would be written on the script almost entirely by the director.

Although the format in Figure 6–2, a film script, is different, note that the terminology and the visual results are virtually the same. The numbers in the left-hand column refer to each *shot* or *sequence*, with film scripts usually shot out of sequence and all scenes in a particular setting done with the cast at that location. The numbers make it possible to easily designate which sequences will be filmed at a given time or on a given day, such as "Living Room Set—sequences 42, 45, 46, 78, 79, 81."

Types of Shots

Shot designations range from the close-up to the medium shot to the long shot. Within these categories there are gradations, such as the medium long shot and the extreme close-up. The writer indicates the kind of shot and the specific subject to be encompassed by that shot. The use of the terms and their meanings apply to both the film and the television format. Here are the most commonly used shots:

Close-Up

This may be designated by the letters *CU*. The writer states in the script "CU Harry" or "CU Harry's fingers as he twists the dial of the safe," or "CU Harry's feet on the pedals of the piano." The close-up of the immediate person of a human subject will usually consist of just the face and may include some of the upper part of the body, with emphasis on the face, unless specifically designated otherwise. The letters XCU or ECU stand for extreme close-up and designate the face alone. The term shoulder shot indicates an area encompassing the shoulders to the top of the head. Other designations are bust shot, waist shot, hip shot and knee shot.

Medium Shot

This may be designated by the letters *MS*. The camera picks up a good part of the individual or group subject, the subject usually

FADE IN

1. EXT. BEACH—SUNRISE—EXTREME LONG SHOT
2. PAN ALONG SHORE LINE AS WAVES BREAK ON SAND
3. EXT. BEACH—LONG SHOT
   Two figures are seen in the distance, alone with the vastness of sand and water surrounding them.
4. ZOOM SLOWLY IN UNTIL WE ESTABLISH THAT THE FIGURES ARE A MAN AND A WOMAN.
5. MEDIUM LONG SHOT
   The man and woman are standing by the water's edge, holding hands, staring toward the sea. They are about 60 years old, but their brightness of look and posture make them seem much younger.
6. MEDIUM SHOT—ANOTHER ANGLE ON THEM
   They slowly turn their faces toward each other and kiss.
7. CLOSE SHOT
8. MEDIUM CLOSE SHOT
   Their heads and faces are close, still almost touching.
   GLADYS:
   I did not feel so beautiful when I was 20.
9. CLOSE SHOT—REGINALD
   as he grins
   REGINALD:
   Me neither. But we weren't in love like this when we were 20.
   CUT TO
10. INT. BEACH HOUSE—ENTRANCE HALL—MORNING
    The door opens and Gladys and Reginald walk in, hand-in-hand, laughing.

Figure 6–2 Standard film format.

filling the screen, but usually not in its entirety, and without too much of the physical environment shown.

Long Shot

The writer may state this as *LS.* The long shot is used primarily for establishing shots in which the entire setting, or as much of it as necessary to orient the audience properly, is shown. From the long shot, the camera may move to the medium shot and then to the close-up, creating a dramatic movement from an overall view to the impact of the essence or selective aspect of the situation. Conversely, the camera may move from the intriguing suspense of the extreme close-up to the clarifying broadness of the extreme long shot. Sometimes the writer designates *establishing shot* instead of *long shot.*

Full Shot

This is stated as *FS.* The subject is put on the screen in its entirety. For example, "FS Harry" means that the audience sees Harry from head to toe. "FS family at dinner table" means that the family seated around the dinner table is seen completely.

Control Room Techniques and Editing

The technicians in the control room have various electronic devices for modifying the picture and moving from one picture to another. The technicians in the film editing room have the same capabilities, except that the modifications are done during the editing process, while in live-type videotaped television the modifications are done during the recording of the program. Further modifications can take place when editing the videotape.

The Fade

The *fade-in* consists of bringing in the picture from a black (or blank) screen. The *fade-out* is the taking out of a picture until a black level is reached. The fade is used primarily to indicate a passage of time, and in this function serves much like the curtain or blackout on the legitimate stage. Depending on the sequence of action, a fast fade-in or fade-out or slow fade-out or fade-in may be indicated. The fade-in is used at the beginning of a sequence, the fade-out at the end. The fade sometimes also is used to indicate a change of place.

The Dissolve

While one picture is being reduced to black level, the other picture is being brought in from black level, one picture smoothly dissolving into the next. The *dissolve* is used primarily to indicate a change of place, but is used sometimes to indicate a change of time. There are various modifications of the dissolve. An important one is the matched dissolve, in which two similar or identical subjects are placed one over the other and the fading out of one and the fading in of the other shows a metamorphosis taking place. The dissolving from a newly lit candle into a candle burned down would be a use of the matched dissolve.

The Cut

The *cut* is the technique most commonly used. It consists simply of switching instantaneously from one picture to another. Care must be taken to avoid too much cutting, and to make certain that the cutting is consistent with the mood, rhythm, pace, and psychological approach of the program as a whole.

The Superimposition

The *super,* as it is sometimes called, means the placing of one image over another, thus creating a fantasy kind of picture. This sometimes is used in the stream-of-consciousness technique when the thing being recalled to memory is pictured on the screen. The superimposition may be used for nondramatic effects very effectively, such as the superimposition of titles over a picture or the superimposition of commercial names or products over the picture.

The Wipe

This is accomplished by one picture literally wiping another picture off the screen in the manner of a window shade being pulled down over a window. The *wipe* may be from any direction: horizontal, vertical, or diagonal. The wipe may also blossom out a picture from the center of a black level or, in reverse, envelop the picture by encompassing it from all its sides. The wipe can be used to designate a change of place or time.

The Split Screen

In the *split screen* the picture on the air is actually divided, with the shots from two or more cameras occupying adjoining places on the screen. A common use is for phone conversations, showing the persons speaking on separate halves of the screen. The screen may be split in many parts and in many shapes.

Key or Matte

A *key* is a two-source special effect where a foreground image is cut into a background image and filled back in with itself. A *matte* is a similar technique, but has the capability of adding color to the foreground image. Character generators (*chirons* or *vidifonts*) electronically cut letters into background pictures. Titles and commercial names of products are keyed or matted. *Chroma key* is an electronic effect that cuts a given color out of a picture and replaces it with another visual. Newcasts use this technique; the blue background behind the newscaster is replaced with a slide or taped sequence.

PUBLIC SERVICE ANNOUNCEMENTS AND COMMERCIALS

Public service announcements (PSAs) and commercials differ mainly in that most of the former promote not-for-profit ideas, products and services—which may include ideological participation through personal time, energy, or money contributions—while the latter sell, for profit, products and services. All commercials are designed to sell something. It is the existence of these commercials that provide the economic base for the American broadcasting system as it is now constituted.

Some commercials are awful because they insult our aesthetic sensibilities. Some are awful because they insult our logic and intelligence. Some are awful because they play on the emotions of those least able to cope with the incitements to buy, such as children.

Some commercials are good because they are, indeed, more aesthetically pleasing than the programs they surround. Some are good because they are educational and provide the viewer with informational guidelines on goods and services.

This puts the writer squarely in the middle. On one hand, the writer has a responsibility to the agency and advertiser, creating not only the most attractive message possible, but one that convinces and sells. On the other hand, the writer has the responsibility of being certain that the commercial has a positive and not a negative effect on public ethics and actions.

The pressures of audiences, civic and citizen organizations, some professions, and the federal government from time to time change the approaches to commercial writing and presentation. At one time some commercials were blatantly racist. Most advertisers and agencies now avoid stereotyped portrayals of minorities; some include minorities as nonstereotyped characters in commercials. Prejudicial portrayals are not limited to race. Negative ethnic references, particularly those related to national origins, have been criticized. Probably the most flagrant area of stereotyping is that regarding women.

Lengths

Spot announcements may be commercial or noncommercial. The noncommercial kind are called PSAs. Commercial spots may be inserted either within the course of a program or during the station break.

Spots may be of various lengths. The overwhelming number are thirty seconds long, although when the economy is up there is a marked increase in sixty-second spots. Recently, with advertising costs climbing, the *split-30* or 15-second spot has become popular. On cable TV two-minute and longer spots are seen frequently.

Public Service Announcements

Stations usually receive PSAs already written and produced for the distributing organization. Service groups, government agencies, and other organizations devoted to activity related to the public welfare, such as public health departments, educational associations, societies aiding the handicapped, and ecology groups, among others, have devoted more and more time in recent years to special television and radio workshops for their regular personnel and volunteer assistants.

The good PSA is like the good commercial: it puts the product or service in the setting, using the strongest attention-getting, attention-keeping, and persuasive elements, including personalities and drama. Most often, however, the lack of funds keeps PSAs direct and simple, using a personality or an event to attract attention. Figure 6–3 combines actress Meryl Streep and a Mother's

Meryl Streep script—TV—30 seconds

Streep

My baby will never have polio, diphtheria or measles.

We've cured them.

But one of the last childhood major diseases remains. Nuclear War. Deadlier than all the rest combined.

Please join Millions of Moms in sharing information about the prevention of nuclear war.

Send your name and address to MOM, Post Office Box B, Arlington, Massachusetts 02174.

You can help cure a major childhood disease.

Figure 6–3 Public Service Announcement. (Courtesy of Women's Action for Nuclear Disarmament Education Fund.)

Day campaign by the Women's Action for Nuclear Disarmament Education Fund.

Techniques of Writing Commercials

Barbara Allen, television writer and producer, sets forth five preliminary steps in putting commercials together: (1) know the product or service; (2) pick the central selling idea; (3) choose the basic appeal; (4) select the format; and (5) start writing.

Emotional Appeals

The appeal of the commercial is an emotional one. By emotional we do not mean the evoking of laughter or tears. Emotional appeal means, here, the appeal to the nonintellectual, nonlogical aspects of the prospective customer's personality. It is an appeal to the audience's basic needs or wants. There are a number of basic emotional appeals that have been particularly successful and upon which the writer of commercials may draw as the motivating factor within any individual commercial. The appeal to self-preservation is perhaps the strongest of all. Drug commercials, among others, make good use of this appeal. Another strong appeal is love of family, as evidenced by any insurance commercial. Other widely used emotional appeals include patriotism, good taste, reputation, religion, loyalty to a group, and conformity to public opinion.

Audience Analysis

Before choosing and applying the specific emotional or logical appeals, the writer must know, as fully as possible, the nature of the audience to whom the message is directed. In television it is often impossible to determine many specifics about a given audience at a given air time. The audience usually is a disunified mass of many attitudes and interests, economic, social, political, and religious levels, spread out over a broad geographical area.

There are some basic elements of audience analysis, called *demographics,* that the writer may apply. These are age, sex, size, economic level, political orientation, primary interests, occupation, fixed attitudes or beliefs, educational level, ethnic background, geographical concentration, and knowledge of the product. The writer should try to include appeals to all the major groups expected to watch the given program—and commercial. The writer should be careful, however, not to spread the message too thin.

Organization of the Commercial

Inasmuch as the commercial's primary purpose is to persuade, the writer should be aware of the five basic steps in persuasive technique. First, the commercial should get the attention of the audience. This may be accomplished by many means, including humor, a startling statement or picture, a rhetorical question, a vivid description, a novel situation, or a suspenseful conflict. Sound, specifically the use of pings, chords, and other effects, effectively attracts attention, too.

Second, after attention is obtained, the audience's interest must be held. Following up the initial element with effective examples, testimonials, anecdotes, statistics, and other devices, visual or aural, should retain the audience's interest.

Third, the commercial should create an impression that a problem of some sort exists, related to the function of the product advertised. After such an impression has been made, then, fourth, the commercial should plant the idea in the audience's mind that the problem can be solved by use of the particular product. It is at this point that the product is "sold," the fifth step in persuasion.

Writing Styles

The writer constantly must be aware of the necessity for keeping the commercial in good taste. Usually, the writer will avoid slang and colloquialisms unless they have specific purposes in the commercial. The writer should be certain that the writing is grammatically correct. Action verbs are extremely effective, as are concrete, specific words and ideas. If an important point is to be emphasized, the writer must be certain to repeat that point in the commercial, although in different words or in different forms. One exception would be the presentation of a slogan or trade mark; in this case word for word repetition is important. The writer should avoid, if possible, the use of superlatives, false claims, phony testimonials, and other elements of obvious exaggeration.

The Television Storyboard

Commercial producers (and account executives and sponsors) like to see as fully as possible the visual contents for a prospective commercial in its early stages. For this purpose a *storyboard* is used. The storyboard usually is a series of rough drawings showing the sequence of picture action, optical effects, settings, and camera angles, and it contains captions indicating the dialogue, sound and music to be heard. There are frequently many refinements from the storyboard that sells the commercial to the advertiser, to the finished film or tape that sells the product to the viewer. Some producers work from storyboards alone. Others want scripts containing the visual and audio directions and dialogue. Figures 6–4 and 6–5 show the storyboard and the photos of the aired final product of an award-winning commercial.

Commercial Formats

There are five major format types for commercials: straight sell, testimonial, humorous, musical, and dramatization. Any single commercial may consist of a combination of two or more of these techniques.

Straight Sell

This should be a clear, simple statement about the product. The *straight sell* may hit hard, but not over the head and not so hard that it may antagonize the potential customer. The straight sell is straightforward, and although the statement about the product is basically simple and clear, the writing technique sometimes stresses a *gimmick,* usually emphasizing something special about the product, real or implied, that makes it different or extra or better than the competing product. A slogan frequently characterizes this special attribute.

The Testimonial

The *testimonial* commercial is very effective when properly used. When the testimonial is

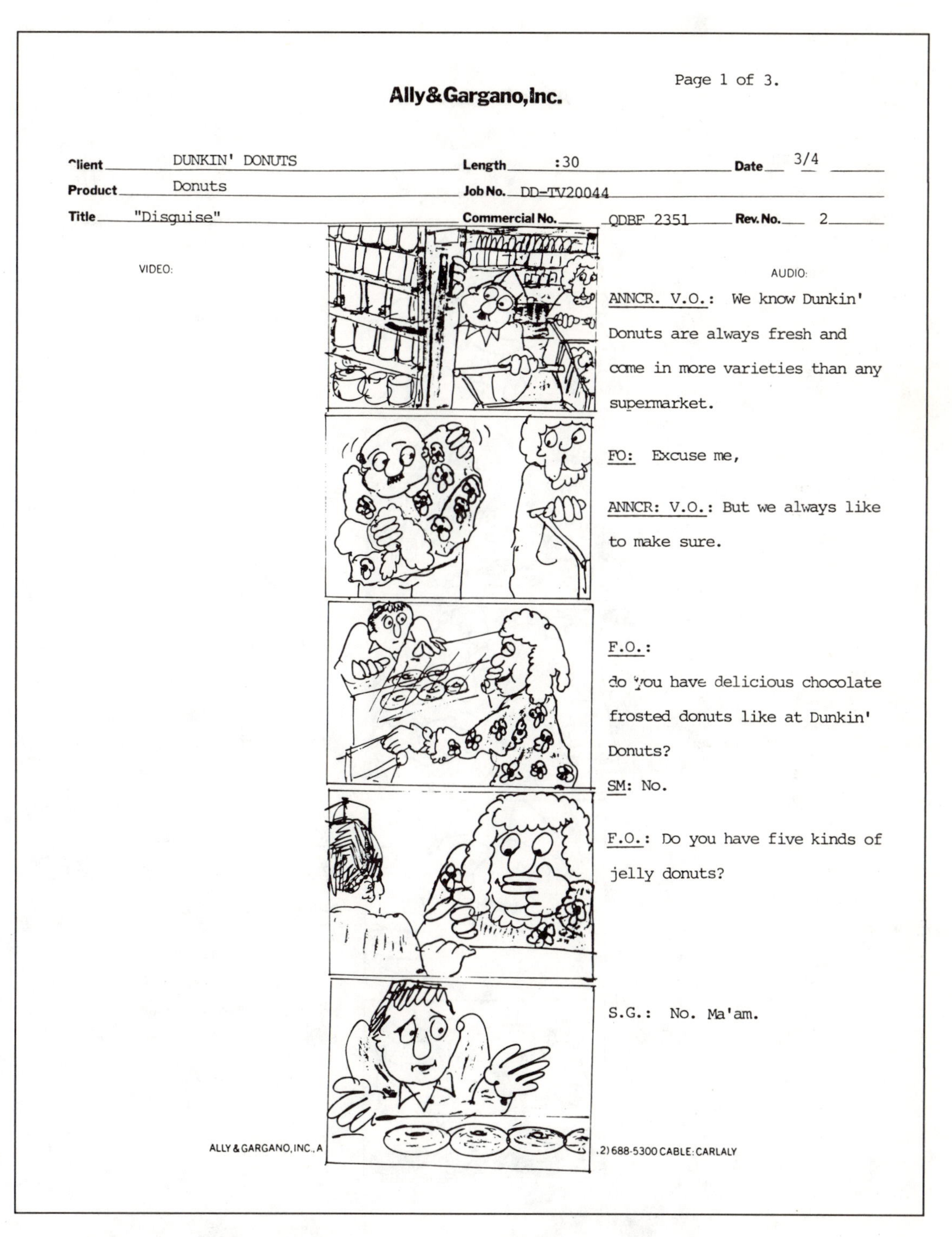

Page 1 of 3.

Ally&Gargano,Inc.

Client DUNKIN' DONUTS Length :30 Date 3/4

Product Donuts Job No. DD-TV20044

Title "Disguise" Commercial No. QDBF 2351 Rev. No. 2

VIDEO:

AUDIO:

ANNCR. V.O.: We know Dunkin' Donuts are always fresh and come in more varieties than any supermarket.

FO: Excuse me,

ANNCR: V.O.: But we always like to make sure.

F.O.:
do you have delicious chocolate frosted donuts like at Dunkin' Donuts?

SM: No.

F.O.: Do you have five kinds of jelly donuts?

S.G.: No. Ma'am.

ALLY & GARGANO, INC., A .2) 688-5300 CABLE: CARLALY

Figure 6–4 Three-page storyboard for a commercial. (Courtesy of Dunkin' Donuts of America, Inc.)

continued

Page 2 of 3.

Ally&Gargano,Inc.

Client DUNKIN' DONUTS Length :30 Date 3/4

Product Donuts Job No. DD-TV20044

Title "Disguise" Commercial No. QDBF 2351 Rev. No. 2

VIDEO: AUDIO:

F.O.: Were these made fresh this morning?

(GROUP OF WORKERS STARE AT HIM)

SM: No.

F.O.: Good.

V.O.: Dunkin' Donuts. Up to 52 varieties made fresh day and night.

No supermarket can say that.

ALLY & GARGANO, INC., AD ... 2) 688-5300 CABLE: CARLALY

Figure 6–4 *continued*

Page 3 of 3

Ally&Gargano,Inc.

Client DUNKIN' DONUTS Length :30 Date 3/4
Product Donuts Job No. DD-TV20044
Title "Disguise" Commercial No. ODBF 2351 Rev. No. 2

VIDEO:

AUDIO:

SUPER:

DUNKIN' DONUTS. IT'S WORTH THE TRIP.

DUNKIN DONUTS

ALLY & GARGANO, INC., ADVERTISING, 805 THIRD AVENUE, NEW YORK, N.Y. 10022 (212) 688-5300 CABLE: CARLALY

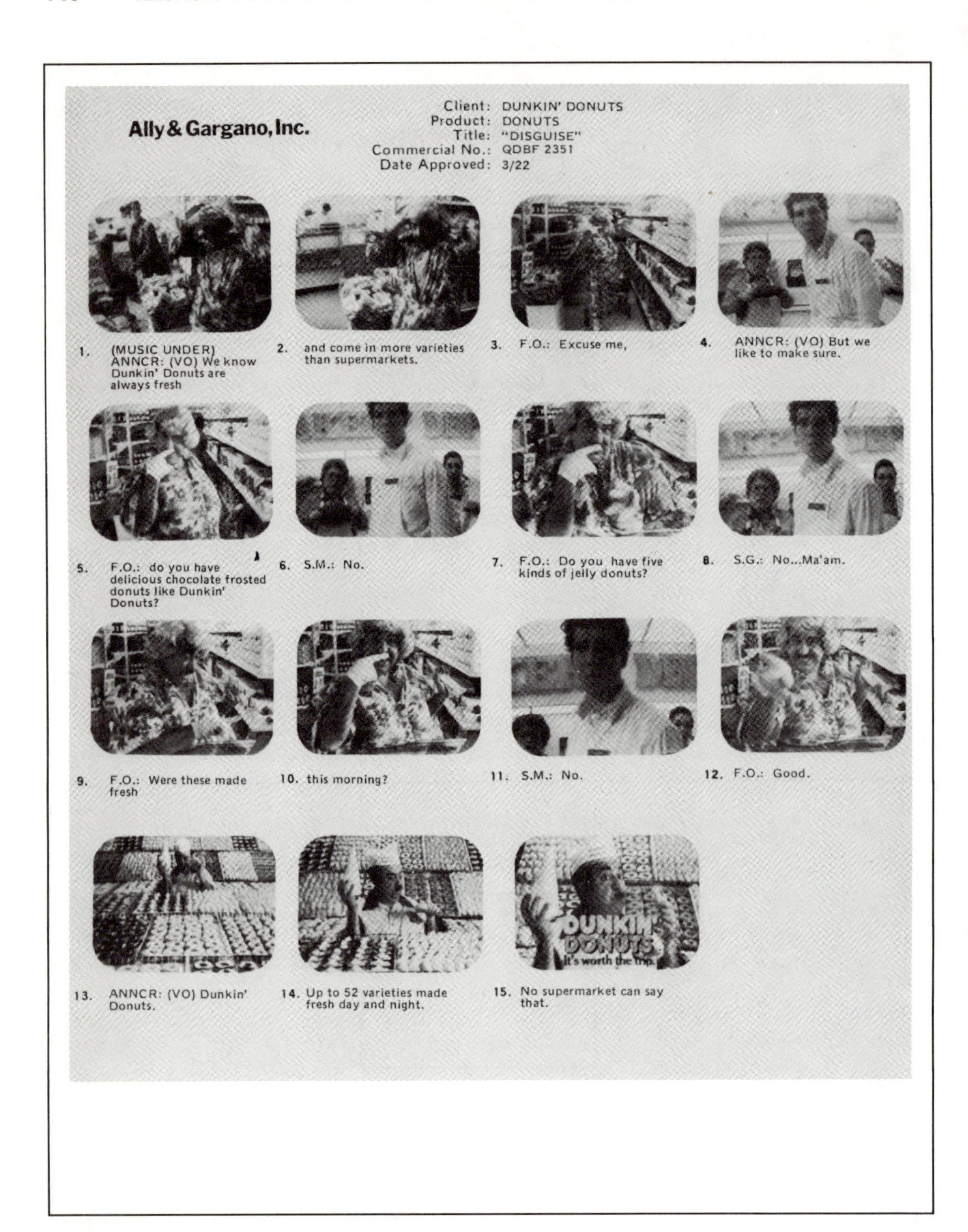

Figure 6–5 Photoboard for same commercial as in Figure 6–4. (Courtesy of Dunkin' Donuts of America, Inc.)

given by a celebrity—whose social and economic status is likely to be quite a bit higher than that of the average viewer—the emotional appeals of prestige, power, and good taste are primary. What simpler way to reach the status of the celebrity, if only in one respect, than by using the same product he or she uses? The writer must make certain that the script fits the personality of the person giving the endorsement.

An alternative to the traditional celebrity testimonial is the testimonial from the average man or woman: the worker, the homemaker, the man- or woman-in-the-street with whom the viewer at home can more easily identify.

Humor

Just as public attitudes toward humor change over the years, so do the humorous approaches in commercials. Always an effective attention-getter, *humor* in commercials, to be successful, must reflect the humorous trends of the time. Satire and parody mark some of the most successful humorous commercials. The most effective parodies have been those which have a story line, even a limited one, in which the situation is dominant. Within the situation are the references to the product.

Music

The *musical* commercial has always been one of the most effective for having an audience remember the product. How many times have you listened to a song on radio or television, been caught up in its cadence, and then suddenly realized it was a commercial and not the latest popular hit tune? Music has been so effective in writing commercials that many of us have come to identify and remember Coca-Cola, McDonald's, and United Air Lines, among others, first with their theme music and only secondarily with a particular sales message. In the mid-1980s, as music videos gained in popularity, musical commercials began to use MTV scenic, dance, prop, electronic, and sound techniques for their formats.

Dramatizations

A *dramatization* is, in effect, a short play—a happening that creates suspense and reaches a climax. The climax is, of course, the revelation of the attributes of the product. In the classic structure of the play form, the desired resolution is the members of the audience all rushing out of their homes to buy the particular product. Dramatizations frequently combine elements of the other major commercial forms, particularly music, testimonials, and humor.

NEWS

Any real happening that may have an interest for or effect upon people is news. The television reporter has a limitless field. Anything from a cat up a tree to the outbreak of a war may be worthy of transmission to the mass media audience. The gathering of news, however, is not our primary concern here. The writing of news broadcasts is.

The two major agencies, the Associated Press (AP) and United Press International (UPI), which serve as news sources for the newspapers, also service television and radio stations. The larger networks have their own news gathering and reporting organizations.

The news usually is prepared by writers in a special news department. Some small stations do not have special news programs or departments, so news broadcasts are prepared by available personnel. In recent years, with the development of lightweight portable minicam equipment (also called ENG), local news has become as important to most stations as network news.

Styles of Writing

The writer of the news broadcast is, first of all, a reporter whose primary duty is conveying the news. The writer, in fact, frequently is the reporter who presents the story on the air. The traditional *five Ws* of news reporting apply. In its limited time, the television report should tell what, where, when, who and, if possible, why. The key word is *condensation.* The writer must be aware of the organization of the broadcast in order to provide the proper transitions, which should be clear and smooth between each story. The writer should also be aware of the content approach, whether it is straight news, analysis, or personal opinion, so as not to confuse editorializing with news.

A writer should consider the time of the day the broadcast is being presented; whether the audience is at the dinner table, seated comfortably in the living room, or rushing madly to get to work on time. The writer should think of the news as dramatic action. The story with an obvious conflict (environmental pollution, arms control, budget deficits, a political contest, the baseball pennant race) attracts immediate attention. Because action is important, the stories should be written with verbs. The immediacy of television, as opposed to the relatively greater time lapse between the occurrence and reporting of the incident in print journalism, permits the use of the present tense in stories about events that happened within a few hours preceding the newscast. Obviously, the writer is dependent upon the visuals available for any given story, whether on tape, a photograph, or live action being covered by an ENG crew.

The writer should begin the news story with precise, clear information. The opening sentence should be, if possible, a summary of the story as a whole. A writer must not include too many details and should remember that the audience hears the news only once and, unlike the newspaper reader, cannot go back to clarify particular points in the story. The audience must grasp the entire story the first time it hears it. The writing, therefore, must be simple and understandable and, without talking down to the audience, colloquial in form. This does not imply the use of slang or illiterate expressions, but suggests informality and understandability. Repetition must be avoided, and abstract expressions and words with double meanings should not be used. The information should be accurate and there should be no possibility of a misunderstanding of any news item.

News Broadcast Types and Organization

The most common type of television news broadcast is the straight news presentation, usually in half-hour periods. There are also commentator-personalities who present news analysis and/or personal opinions on the news, frequently included in the straight news show. Networks and stations sometimes have *specials* or "minidocumentaries" that probe the news.

Most news programs are a combination of the live announcer or announcers, film, and videotape and, in some cases where a story is breaking at that moment, live remotes. There can be multiple pickups from various studios or sites, involving reporters closer to the scenes of the events. Networks frequently present news roundups from various parts of the country and, through satellite, from various parts of the world.

In the straight news program the writer should look for a clear and logical organization, no matter what the topic or approach. One such organization is for the placement of stories to follow a topical order; that is, the grouping of similar stories into sections, although the order of the sections themselves may be an arbitrary one. A geographical grouping and order is another organizational form. For example, the news coverage may move from North America to Europe to Asia to South America to Africa to the rest of the world. Another frequently used grouping organizes the material into international, national, and local news categories. The most common approach is to place the most important story first in order to get and hold the audience's attention, much as does the lead story in the newspaper.

The organization is determined, in part, by the audience being reached. In the midmorning newscast, for example, stories sometimes are chosen and placed to appeal primarily to women, the bulk of the listening group at that time. In the early evening the organization sometimes is one that will reach most effectively the viewer who has just returned from work. The news broadcast just before prime time on TV frequently seeks to reach a family group watching together. The time of day is also important in relation to what the audience already knows of the news. In the early morning newscast it is desirable to review the previous day's important late stories. In the late evening broadcast the current day's news should be reviewed and the audience should be prepared for the next day's possible happenings.

The physical format of the news show may vary. It may begin with an anchorperson giving the headlines, followed by a commercial, and then a correspondent coming in with the details. It may be a roundup of different reporters in different geographical areas giving on-the-scene reports.

Rewriting

One of the newswriter's duties, especially on the local level, is rewriting or adapting stories to include a local angle. For example, segments in a topical grouping of stories dealing with the economy might be rewritten to reflect their relationship to the local unemployment figures, with an ENG crew sent out to get visuals on the topic.

Perhaps the most common form of rewriting is updating. An important story doesn't disappear after it is used once. Yet, to use exactly the same story in both the 6:00 P.M. and 11:00 P.M. newscasts (depending on the time zone) is likely to turn off those listeners who hear it more than once and conclude that the station is carrying stale news. There are several major areas to look for in updating news stories. First, the writer determines if there is any further hard news, factual information to add to the story. Second, if the story is important enough it is likely that investigative reporting will have dug up some additional background information not available when the story was first broadcast. Third, depending on the story's impact on society, it will have been commented upon after its initial release by any number of people from VIPs to ordinary citizens. In addition, a story may by its very nature relate to other events of the day, that relationship being made clear in the rewriting.

Process

Remember that television is visual. The reporter, feeding a remote or in the studio, must have a visual personality. The television news personalities have become the stars of the medium. The physical setting should be interesting and attractive, consistent with the concept of informational and exciting news. Even the presentation of content that in itself may be undramatic should be visually stimulating. For example, weather reporters use different techniques to report the nightly weather segment of the news program. Television news should stress the visual and may use videotape, live remotes, film, slides, photographs, inanimate objects, and, where necessary for emphasis or exploration in depth, even guests in the studio. Except for relatively extensive use of an anchorperson to keep together the physical continuity of the program or, through that person's special personality, to instill confidence or, as a star, to motivate viewing by the audience, the TV news program should show the news, not tell about it.

Writing the script that appears over the air is only the final stage of a long, arduous, and frequently complicated process. Planning and development begin early in the day, for both network and local news. As early as possible, assignments of reporters are made, in the network situation in literally all parts of the world, in the local situation to standard beats (such as city hall, the police station) and to special events that may occur that day. Based on anticipated news stories, a writer prepares a *rundown* sheet, that is, a preliminary outline of who will cover what and where. Later on, as some stories are already on tape and other breaking events are being covered, a writer can prepare a more detailed outline, including anticipated times for each segment of news to be used. Finally, as air time approaches, the final script is prepared, with the writer making sure that there is sufficient and smooth continuity for the studio announce team to use between segments of tape, film, and live remotes that already have narration or interviews. A network news show may have as many as nine or ten steps in this process; a local news show may have only a few.

The final scripts are very much alike from station to station. Usually they use the standard two-column format, the visual information on the left (letting the studio anchors know what's coming and clarifying for the director and technical director what visual resource to insert), and the audio information on the right (including the continuity to be read by the anchors and other instructions to them). A typical example is provided in Figure 6–6, from the final script for a 10:00 P.M. local news show on a local UHF station.

Figure 6–6 Local television news script. (Courtesy of WLVI-TV, Boston, MA.)

VIDEO	AUDIO
(WIPE) ENG/SIL/VO HYNES	AS SEABROOK GETS CLOSER TO GOING ON-LINE, THERE'S MORE CRITICISM ABOUT THE DESIGN OF THE PILGRIM NUCLEAR PLANT IN PLYMOUTH.
KEY-FILE	ROBERT POLLARD, A FORMER N-R-C OFFICIAL NOW WITH THE UNION OF CONCERNED SCIENTISTS, SAYS THE AGENCY IS COVERING UP DESIGN FLAWS AT MANY PLANTS, INCLUDING PILGRIM.
(BUTTED) ENG/SOT	
KEY-POLLARD SOT RUNS :22 OUT— "OF A MAJOR ACCIDENT"	
HYNES ON CAM	IN A RELATED DEVELOPMENT, THE STATE DEPARTMENT OF PUBLIC UTILITIES TODAY BLASTED THE MANAGEMENT OF BOSTON EDISON, WHICH OPERATES THE PILGRIM PLANT. THE D-P-U REJECTED THE COMPANY'S PROPOSED 35-MILLION DOLLAR RATE INCREASE, SAYING EDISON TRIES TO PUT ALL RISKS ON CUSTOMERS INSTEAD OF SHAREHOLDERS.

VIDEO	AUDIO
	IN OTHER NEWS . . .
(BOX-FF-SWITZLER)	REPUBLICAN ROYAL SWITZLER IS STILL RUNNING FOR GOVERNOR TONIGHT. BUT THERE ARE INDICATIONS THAT HE'S CONSIDERING DROPPING OUT OF THE RACE. THIS FOLLOWS DISCLOSURES THAT HE EMBELLISHED HIS MILITARY RECORD.
	SWITZLER HAS CANCELLED ALL CAMPAIGN APPEARANCES WHILE HE MEETS WITH HIS ADVISORS AND G-O-P LEADERS. HE SAYS HE WON'T HAVE ANYTHING MORE TO SAY UNTIL SOMETIME NEXT WEEK.
(WIPE) ENG/SIL/VO KEY-FILE	THERE ARE MORE PROBLEMS TONIGHT FOR SWITZLER'S OPPONENT, GREG HYATT. THE DEMOCRATIC STATE COMMITTEE TODAY FILED CHALLENGES TO SOME OF THE SIGNATURES HYATT GATHERED TO GET ON THE SEPTEMBER PRIMARY BALLOT. THE PROBLEM INCLUDES DUPLICATE SIGNATURES AND SIGNATURES OF UNREGISTERED VOTERS.
HYNES ON CAM	SOME NAMES GATHERED BY SIXTH DISTRICT REPUBLICAN CANDIDATE WILLIAM GOTHA WILL ALSO BE CHALLENGED.
	IN OTHER NEWS TONIGHT . . .
ALISON AND KEY	THE FIRST STEP HAS BEEN TAKEN TO RESOLVE DIFFERENCES BETWEEN MAYOR FLYNN AND BOSTON SCHOOL SUPERINTENDENT LAVAL WILSON.
	THEY HAD A 2-HOUR ONE-ON-ONE MEETING TODAY. THIS FOLLOWED PUBLISHED REPORTS WHERE WILSON CALLED THE MAYOR'S COMPLAINT ABOUT THE SCHOOL SYSTEM "A PUBLICITY GIMMICK."
	TONIGHT, I ASKED WILSON IF FLYNN WOULD BE LEAVING HIM ALONE FOR A WHILE.

continued

VIDEO	AUDIO
ENG/SOT KEY-WILSON SOT RUNS :35 OUT-"A FINE ONE."	
ALISON ON CAM	MAYOR FLYNN WAS UNAVAILABLE BUT HIS EDUCATION ADVISOR FELIX ARROYO SAID FLYNN AND WILSON ARE BOTH WORKING TOWARD THE SAME GOAL—IMPROVING THE BOSTON SCHOOLS.
BOX F/F WESTFIELD KEY ANN	SEVERAL NEW DEVELOPMENTS TONIGHT IN THE PROBE OF ALLEGED SEXUAL MIS-CONDUCT AT WESTFIELD STATE COLLEGE.
ENG/SIL/VO ANN KEY-LAST THURSDAY	INVESTIGATORS FROM THE HOUSE POST-AUDIT AND OVERSIGHT COMMITTEE ARRIVED ON CAMPUS TODAY TO BEGIN THEIR INVESTIGATION.
	FORMER SUPREME COURT JUDGE RUDOLPH PIERCE WAS APPOINTED BY THE STATE BOARD OF REGENTS TO OVERSEE A SEPARATE PROBE. AND STATE ADMINISTRATION AND FINANCE SECRETARY FRANK KEEFE TODAY SENT A LETTER TO COLLEGE OFFICIALS, DEMANDING THEY RETURN A CONTROVERSIAL 10-THOUSAND DOLLAR PAYMENT.
ANN ON CAM	THAT PAYMENT WAS MADE LAST MONTH TO A STUDENT WHO CLAIMED HE WAS SEXUALLY ASSAULTED BY FORMER WESTFIELD STATE PRESIDENT FRANCIS PILECKI . .
BOX-FF TYLENOL KEY HYNES	ANOTHER SCARE INVOLVING TAINTED TYLENOL TONIGHT. BUT HEALTH OFFICIALS SAY THEY KNOW WHAT CAUSED THE PROBLEM.
ENG/SIL/VO HYNES KEY-FILE	A SOUTHBRIDGE WOMAN SAYS SHE FOUND A POWDERY SUBSTANCE IN A BOX OF STANDARD-STRENGTH TYLENOL TABLETS.
	THAT RESULTED IN THE PRODUCT BEING PULLED FROM STORE SHELVES IN THE CENTRAL MASSACHUSETTS COMMUNITY. THE WHITE SUBSTANCE IS AN ANTI-

Figure 6–6 *continued*

VIDEO	AUDIO
	BIOTIC CALLED CEPHA-LEXIN, AND IS MADE BY THE SAME COMPANY THAT PRODUCES TYLENOL. APPARENTLY, THE DRUG GOT INTO A BOTTLE OF TYLENOL DURING MANUFACTURING. LAB TESTS BY THE FOOD & DRUG ADMINISTRATION SHOW NO EVIDENCE OF TAMPERING.
HYNES AND KEY	AN EAST BOSTON MAN IS UNDER GUARD AT MASSACHUSETTS GENERAL HOSPITAL TONIGHT, AFTER BEING SHOT BY A BOSTON POLICE OFFICER.
ENG/WSOT/VO HYNES KEY-3:30 am/EAST BOSTON	35-YEAR OLD DANIEL TRAINOR REPORTEDLY TRIED TO RUN OVER OFFICER JAMES EARLE WITH A STOLEN CAR EARLY THIS MORNING. TRAINOR ATTEMPTED TO FLEE AFTER CRASHING INTO EARLE'S VEHICLE. THAT'S WHEN 2 OTHER POLICE OFFICERS FIRED SEVERAL SHOTS. ONE OF THEM HIT TRAINOR IN THE CHEST.
TWO SHOT HYNES	WHEN HE RECOVERS, TRAINOR WILL FACE CHARGES OF ASSAULT AND ATTEMPTED MURDER.

Special Events

The special event is usually under the direction of the news department and is essentially something that is taking place live and is of interest—critical or passing—to the community. It is usually a remote, on-the-spot broadcast. Special events usually originate independently and include such happenings as political conventions, parades, dedications, banquets, awards, and the openings of new films and supermarkets.

Sometimes special events are merely introduced, presented without comment, and, occasionally, summarized or critiqued when over. Sometimes they are accompanied by commentary. The opening and closing material and, frequently, transition and filler material are provided by the writer. The latter two are sometimes handled directly by the broadcaster who is assigned to the event and who presumably is an expert on the subject being covered.

For events other than those that require only a short intro and outro, the writer should collect as much material as possible. News stories, maps, press releases, historical documents, books, photographs, locales, and similar sources can be pertinent and helpful in preparing continuity. Copy should be prepared for all emergencies as well as for opening, closing, transition, and filler uses. Material should include information on the personalities involved, the background of the event, and even on probable or possible happenings during the event. Usually covered as a special event by the news department are political speeches, interviews, and press conferences. Some elements in such presentations overlap the formats of *features* and *talk programs*.

SPORTS

At one time the sports department of a station or network was an offshoot of the news department. But the phenomenal growth of live sports event coverage has given sports new status in broadcasting and more and more sports divisions are separate, independent functions. The smaller the station, of course, the greater the likelihood that sports will be a part of the news department. The writing of sports is similar to the writing of news. If anything, the style for sports broadcasts must be even more precise and more direct than the news broadcasts. The language is more colloquial and, though technical terms are to be avoided so as not to confuse the general audience, the writer of sports may use many more expressions relating to a specialized area than can the writer of news.

Types of Sports Programs

The straight sportscast concentrates on recapitulation of the results of sports events and on news relating to sports in general. Some sportscasts are oriented solely to summaries of results with appropriate tape segments illustrating the highlights of the contest being reported.

The sports feature may include live or taped interviews with sports personalities, anecdotes or dramatizations of events in sports, human interest or background stories, or remotes relating to sports but not in themselves an actual athletic event.

The most popular sports broadcast is, of course, the live athletic contest while it is taking place.

Organization of Sports News

Formats for the sports news broadcast parallel those of the regular news broadcast. The most common approach is to take the top sport of the particular season, give all the results and news of that sport, and work toward the least important sport. In such an organization the most important story of the most important sport is given first unless a special item from another sport overrides it. Within each sport the general pattern in this organization includes the results first, the general news (such as trades, injuries, and so forth) next, and future events last. If the trade or injury is of a star player or the future event is more than routine, such as the signing for a championship fight, then it will become the lead story. The local sports scene is usually coordinated with the national sports news, fitting into the national reporting breakdowns, but coming first and given more emphasis in the local broadcast.

Sports Special Events

The live on-the-spot coverage of an athletic contest is the most exciting and most popular sports program. The newspaper and magazine cartoons showing a viewer glued to a television set for seven nights of baseball in the summer or seven nights of football, basketball, and hockey in the winter are no longer exaggerations. The sports special event can be other than a contest, however. Coverage of an awards ceremony, of an old-timer's day, of a Cooperstown Hall of Fame induction, or of a retirement ceremony are all special events that are not live contests on a playing field.

The sports broadcaster must have filler material; that is, information relating to preevent action and color, statistics, form charts, information on the site of the event, on the history of the event, about the participants, human interest stories, and similar materials that either heighten the audience's interest or help clarify the action to the audience. This material must be written and made available to the broadcaster to be used when needed, specifically during lulls in the action, and in pregame and postgame opening and closing segments. Staff researchers and writers usually prepare this material. Sports broadcasters, however, are expected to be experts in their field and to know and to provide much of their own material.

The primary function of the writer for the live contest is that of a researcher and outliner. The script may be little more than an outline and/or a series of statistics, individual unrelated sentences, or short paragraphs with the required background and transition continuity.

FEATURES

The feature falls somewhere between the special event and the documentary. While the special event is coverage of an immediate newsworthy happening, sometimes unanticipated, the feature is preplanned and carefully prepared. There are special events, of course, such as sports events, that are preplanned. The special event usually is live, while the feature usually is filmed, taped, or, if live, produced from a script or at least a routine sheet or detailed rundown sheet. Special events usually are public presentations that stations arrange to cover. Features usually are prepared solely for television and radio presentation and generally are not presented before an in-person audience. The special event is part of the stream of life while the feature is designed by a producing organization.

Features usually are short: two to five minutes for fillers and fifteen to thirty minutes for full programs of a public service nature. The subject matter for the feature varies. Some sample types would be the presentation of the work of a special service group in the community, a story on the operation of the local fire department, an examination of the problems of the school board, a how-to-do-it broadcast, or a behind-the-scenes story on any subject—from raising chickens to electing public officials.

Writing Approach

Because it comes close to the documentary, the feature requires careful research, analysis, and evaluation of material, and writing based on detail and depth. That does not mean that it requires a full and complete script. Because the feature is composed, frequently, of a number of diverse program types, such as the documentary, the interview, the panel discussion, and the speech, it may be written in routine sheet or rundown form. Some features have combinations of script, rundown sheet, and routine sheet.

Because the feature is usually a public service presentation it often contains informational and educational content. But it doesn't have to be purely factual or academic in nature. It can even take the form of a variety show or a drama, or certainly have elements of these forms within the program as a whole. The feature is an eclectic form and can be oriented around a person, an organization, a thing, a situation, a problem, or an idea.

THE DOCUMENTARY

A good documentary combines news, special events, features, music, and drama. Some consider it the highest form of the television art. A good documentary not only synthesizes the creative arts of the broadcast media, but makes a signal contribution to public understanding by interpreting the past, analyzing the present, or anticipating the future. A good documentary is not afraid to deal in depth with a controversial subject.

Form

Although the documentary is dramatic, it is not a drama in the sense of the fictional play. It is more or less a faithful representation of a true story. This is not to say, however, that all documentaries are unimpeachably true.

Editing and narration can make any series of sequences seem other than what they really are. The semidocumentary or docudrama has achieved a certain degree of popularity. Based on reality, it is not necessarily factual. It may take authentic characters but fictionalize the events of their lives; it may present the events accurately but fictionalize the characters; it may take real people and/or real events and speculate, as authentically as possible, in order to fill in documentary gaps; it may take several situations and characters from life and create a semitrue composite picture.

Although the documentary deals with issues, people, and events of the news, it is not a news story. It is an exploration behind and beneath the obvious. It goes much more in depth than does a news story, exploring not only what happened, but, as far as possible, the reasons for what happened, the attitudes and feelings of the people involved, the interpretations of experts, the reactions of other citizens who might be affected, and the implications and significance of the subject not only for some individuals, but for the whole of society.

Procedure

Essentially, the documentary contains the real words and actions of real persons and recordings or live action of real events. If there are no visual or aural records available, writings (published and unpublished), letters, photos, drawings, and interviews with people who knew the human subjects or witnessed the events are used. These materials must be put together into a dramatic, cohesive whole and edited according to an outline, then a script.

First, the writer must have an idea. What subject of public interest is worthy of documentary treatment? The idea for the program frequently comes not from the writer, but from the producer or other sources. All documentaries should have a point of view. What is the purpose of the particular documentary? When the subject and point of view are determined, the real work starts; from a tentative outline, to thorough research in libraries, to personal visits to people and places, to investigations of what video and audio materials are already available on the subject. When the research is completed, the writer can prepare a more definitive outline.

After all the materials have been gathered and reviewed, usually many times, the development of a final script can begin. The final script is used for the selection and organization of the specific materials to be used in the final taping/filming and editing of the program. It is not surprising, considering the high degree of coordination and cooperation needed to complete a good documentary, that in many instances the writer also serves as producer and even as director and/or editor.

The picture may be the primary element in any given sequence in the TV documentary, with the narration and taped dialogue secondary. The people and their actions may actually be seen and thus understood, rather than being imagined through verbal descriptions of what they did and saw. On the other hand, the words of the people and the narrator may be the prime movers, with the pictures merely filling in visually what is being described in words. The key to good documentary writing is human interest.

The resources of a major network are not necessary to produce a first-rate television documentary. With careful planning and imagination local stations can produce dramatic and pertinent documentaries. Though not usually controversial, the subject of libraries is a significant one in most communities, and Barbara Allen decided that it was pertinent to viewers in the many cities within the coverage of her station, WGAL, Lancaster, Pennsylvania. Her approach was to take an ostensibly inanimate thing and humanize it. In doing so she captured many of the aspects of libraries that relate to human drama—in this case those of a worrisome nature that require action on the part of the viewers. Her approach was to dramatize the problem, but in the form of a factual statement, not a semidocumentary. A portion of her script is shown in Figure 6–7.

The popularity of the network documentary-type program "60 Minutes" and the development of ENG equipment promoted the growth of minidocumentaries in the 1980s,

Figure 6–7 Detailed feature or documentary outline. (Courtesy of Barbara Allen.)

LIBRARIES: BRUISED, BATTERED AND BOUND

VIDEO	AUDIO
CU OF INITIALS CARVED IN TABLES, WALLS, ETC., FOR EACH LOCATION	MUSIC UNDER—LOVE THEME FROM "ROMEO AND JULIET"
	BARB: This is a love story with an unhappy ending. In Harrisburg, R.P. loves B.L. In Lebanon, A.M. loves P.S. In York, it's M.O. and S.T. In Reading, C.K. loves P.R. and in Lancaster, Brenda loves Bill. START TO FADE MUSIC
COVER SHOT OF TABLE TOP	But love is a very private relationship and these initials are written in very public places. MUSIC OUT
SUPER TITLE SLIDE OVER TABLETOP	They are your public libraries and they are Bruised, Battered and Bound.
DISSOLVE TO COVER OF BARB AND LIBRARIANS AT TABLE ZOOM IN TO BARB	Hello, I'm Barbara Allen. With me around this bruised and battered library table are five librarians from the Channel 8 area. They're not here to tell you about what your local library has to offer. They're here to talk about larceny, decay, suffocation and rape. These things are happening in your library right now. If you don't stop them, the next time you visit your library, you may be greeted by this.
:05 FILM PERSON PUTTING CLOSED SIGN IN WINDOW	
BARB, THEN MR. DOHERTY	(INTRODUCE MR. DOHERTY, CHAT WITH HIM ABOUT CLOSED SIGN AT READING PUBLIC LIBRARY AND ASK HIM ABOUT THE PROBLEMS AT THE READING LIBRARY THAT YOU CAN SEE)

continued

VIDEO	AUDIO
1:15 FILM SHOWING EXTERIOR OF LIBRARY AND VISUAL PROBLEMS INSIDE	MR. DOHERTY VOICE OVER FILM
MR. DOHERTY	(CHAT WITH BARB ABOUT ONE PROBLEM YOU CAN'T SEE)
BARB, THEN MISS YEAGLEY	(ASK ABOUT PROBLEMS YOU CAN SEE AT MARTIN MEMORIAL LIBRARY, YORK)
1:15 FILM SHOWING EXTERIOR OF LIBRARY AND VISUAL PROBLEMS INSIDE	MISS YEAGLEY VOICE OVER FILM
MISS YEAGLEY	(CHAT WITH BARB ABOUT ONE PROBLEM YOU CAN'T SEE)
BARB INTRODUCES MR. GROSS	(ASK ABOUT PROBLEMS YOU CAN SEE AT THE HARRISBURG PUBLIC LIBRARY)
1:15 FILM SHOWING EXTERIOR OF HARRISBURG LIBRARY AND VISUAL PROBLEMS INSIDE	MR. GROSS VOICE OVER FILM
MR. GROSS	(CHAT WITH BARB ABOUT ONE PROBLEM YOU CAN'T SEE)
BARB INTRODUCES MR. MARKS	(ASK ABOUT PROBLEMS YOU CAN SEE AT LEBANON COMMUNITY LIBRARY)

Figure 6–7 *continued*

fitting into magazine-format programs in local stations or in syndication, usually in the early evening. The minidocumentary follows the same approach as "60 Minutes," but instead of each segment being from twelve to twenty minutes in length, the individual stories run from some five to seven minutes. Figure 6–8 shows the opening and closing segments of one such minidocumentary, "The New Right," written and produced by Ron Blau for "Chronicle." The script is reproduced, typos and all, to illustrate the changes that take place from prepared script to final working script.

TALK PROGRAMS

Talk program is used sometimes as an all-inclusive term, encompassing virtually all program types that are not news, documentaries, dramas, features, game shows, music, or commercials. Included in talk programs are interviews, discussions, and speeches. Interview and discussion programs are outlined, either in rundown or routine sheet form. A principal reason they cannot be prepared in complete script form is that the very nature of an interplay of ideas and feelings among people requires extemporaneity. An-

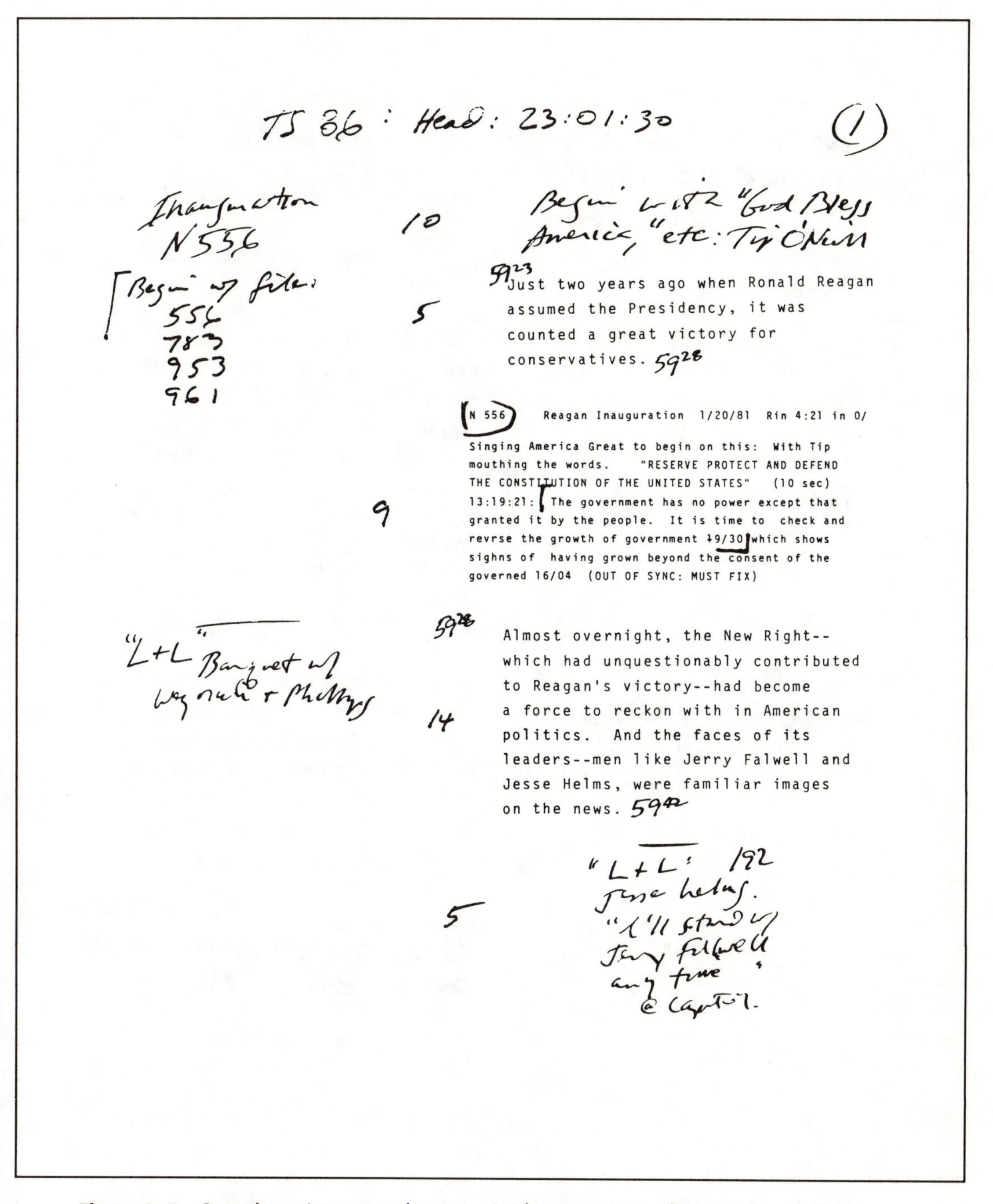

TS 86 : Head: 23:01:30 (1)

Inauguration N556 — 10 — Begin with "God Bless America," etc: Tip O'Neill

[Begin w/ file: 556 783 953 961

5 — 59[23] Just two years ago when Ronald Reagan assumed the Presidency, it was counted a great victory for conservatives. 59[28]

9 — N 556 Reagan Inauguration 1/20/81 Rin 4:21 in 0/
Singing America Great to begin on this: With Tip mouthing the words. "RESERVE PROTECT AND DEFEND THE CONSTITUTION OF THE UNITED STATES" (10 sec)
13:19:21: The government has no power except that granted it by the people. It is time to check and revrse the growth of government 9/30 which shows sighns of having grown beyond the consent of the governed 16/04 (OUT OF SYNC: MUST FIX)

"L+L" Banquet w/ [illegible] + Phillips — 14 — 59[28] Almost overnight, the New Right--which had unquestionably contributed to Reagan's victory--had become a force to reckon with in American politics. And the faces of its leaders--men like Jerry Falwell and Jesse Helms, were familiar images on the news. 59[42]

5 — "L+L": 192 Jesse helms. "I'll stand w/ Jerry Falwell any time" @ Capitol.

Figure 6–8 Sample script pages from a minidocumentary: the opening as it appeared in the prepared script *(above),* the opening in the final working script *(page 160),* and the closing segment of the final script *(page 161).* (Courtesy of WCVB Channel 5, Boston.)

continued

10 Begin with "God Bless America," etc: Tip O'Neill	59[23] Just two years ago when Ronald Reagan assumed the Presidency, it was counted a 5 great victory for conservatives. 59[28]
Inauguration N556	(N 556) Reagan Inauguration 1/20/81 Rin 4:21 in 0/
Begin with file: 556 783 953 961	9 Singing America Great to begin on this: With Tip mouthing the words. "RESERVE PROTECT AND DEFEND THE CONSTITUTION OF THE UNITED STATES" (10 sec) 13:19: 21: The government has no power except that granted it by the people. It is time to check and reverse the growth of government 19/30 which shows signs of having grown beyond the consent of the governed 16/04 (OUT OF SYNC: MUST FIX)
"L+L" Banquet with Wagner and Phillips	59[28] 14 Almost overnight, the New Right — which had unquestionably contributed to Reagan's victory — had become a force to reckon with in American politics. And the faces of its leaders — men like Jerry Falwell and Jesse Helms — were familiar images on the news. 59[42]
	5 "L+L" 192 Jesse Helms "I'll stand with Jerry Falwell any time." @ Capitol.

Figure 6–8 *continued*

The New Right seems to be powerful enough to put the President on the defensive. Recently, Ronald Reagan appears to have gone out of his way to reassure these conservatives. Just one week ago, speaking to a convention of religious broadcasters, he spoke strongly in favor of a Constitutional amendment to return prayer to public schools . . . tax credits for private school tuition . . . and other matters dear to the hearts of the New Right.

DEF for Monday, Jan 31, 1983 (NO TIME CODE numbers for ref only) He's addressing religious broadcasters: Christian Media. Talking in terms of a constitutional amendment returning prayer to public schools. Tax credits for private school tuition, Abortion

23/16 We must go forward with unity of purpose and will 23/20 And let us come together, Christians and Jews, let us pray together, march lobby and mobilize every force we have, so that we can end the tragic taking of unborn childrens lives 23/32

Then more on balancing the budget through using the Bible.

In spite of these overtures, in spite of Reagan's declaring 1983 the Year of the Bible, many observers sense the President is doing a balancing act: making sure not to alienate conservatives too badly . . . while at the same time making sure not to be identified solely with a group so far from center.

11/00 MCU It's true of any Pres 11/02 The Presidency is an office which to be successful has got to command the broad middle ground in pol. And it means that those on one fringe or the other who think they own Pres going to feel a bit alienated from him. It just takes a bit of time in any admin for the recog to dawn on these people that they don't own the President 11/26 other reason is that the participants, excluding the interviewer or moderator, usually are nonprofessionals and cannot memorize or "read" a prepared script without seeming strained and stilted.

Many local interview shows, principally news programs, work with a simple opening and closing for the interviewer and a few key questions, rather than a step-by-step outline and routine sheet. However, a look at the kind of intensive preparation and highly extensive and detailed script materials prepared by the most successful and acclaimed interviewers, such as Barbara Walters, explains why the writer should prepare as much of the script as necessary—whether a detailed routine sheet or a simple outline—for the best possible show. Why take a chance with an unprepared question or sequence when the chances of success are better with prepared material? The rundown sheet is the key for most talk programs. The rundown sheet is a detailed listing of all the sequences in a given program, frequently with the elapsed time, if known, for each item. Rundown and routine sheets sometimes include alternate endings of different lengths so that the extemporaneous nature of the program can be maintained and still end on time through the choice of the proper-length sequence for the final item in the program.

The Interview

The interview may be prepared completely, with a finished script for interviewer and interviewee; it may be oriented around an outline, where the general line of questioning and answering is prepared, but the exact words to be used are extemporaneous; it may be completely unprepared or *ad-lib*. Very rarely are interviews either completely ad-lib or completely scripted. The unprepared interview is too risky, with the interviewee likely to be too garrulous, embarrassing or embarrassed, or just plain dull, and the interviewer likely to be faced with the almost impossible task of organizing, preparing, and thinking of appropriate questions on the spot. The prepared script usually results in a stilted, monotonous presentation except when both the interviewer and interviewee are skilled performers who can make a written line sound extemporaneous, a situation not often likely to occur.

The written material for the extemporaneous interview is the rundown and/or routine sheet, a step-by-step outline of the program that includes a list of questions and probable answers as determined in the preinterview session. Sometimes, of course, the interviewee will not be available for a conference before the show, and the interviewer and staff must guess at the likely answers to their questions, based on thorough research on the interviewee.

In all interviews, prepared, extemporaneous, or ad-lib, the writer prepares at least the opening and closing continuity, introductory material about the interviewee, and for each section of the program, lead-ins and -outs for commercial breaks, an outline of the questions, and, if possible, answers. The closing continuity should be of different lengths in case the program runs shorter or longer than expected. The writer should be certain that the background of the guest is clearly presented. Except where the interviewee is very well known, it is sometimes helpful to begin with questions of a human interest nature so that the audience gets to know something about the personality of the guest before the interview is too far along. Even with a well known guest this sometimes is advisable. In the strictly informational, news-type interview this approach could be distractive, although even in such programs the interviewer frequently asks "personality" questions.

There are three major interview types: the opinion interview, the information interview, and the personality interview. Any given interview can combine elements of all three.

The *opinion interview* concentrates on the beliefs of an individual. However, inasmuch as many of the interviews of this nature are with prominent people, usually experts in their fields, such interviews are not only opinion but, to a great extent, information and even personality types.

The *information interview* usually is of the public service type. The information may be delivered by a relatively unknown figure or by a prominent person in the field.

The *personality interview* is the human interest, feature story. The format of the program may be oriented toward one purpose—to probe, to embarrass, or to flatter—or it may be flexible, combining and interweaving these various facets.

We usually think of the interview as static: two or more people talking at each other. However, even in the simplest question-and-answer process, some visual interest can be injected. The visual movement may be of a subjective nature, with the camera probing the facial expressions and bodily gestures of the interviewee. The visual approach may be broader and more objective, with film or tape or photographs of places, events, or personalities referred to by the interviewee. Because television is visual, the interviewer (and writer-producer-director) must be cautioned about misleading the audience, even unintentionally. One classic story is about the television interviewer who made much in preprogram publicity of a forthcoming interview with a famous stripteaser. Although the audience should have known better, many viewers were quite disappointed that she did not do what she obviously could not do on television.

Format is paramount. Each interview program has its own organization and the writer must write for that particular format. Some interview shows open with an introduction of the program, introduce the guest, and then go into the actual interview. Others open cold, with the interview already under way, in order to immediately grab and hold the audience's attention, subsequently cutting in with the standard introductory material.

Interview shows frequently combine various other format elements, including discussion, audience participation, entertainment, and documentary techniques. Their orientations vary, some of them more variety than interview. In the 1980s the "Donahue" and "Oprah Winfrey" programs are illustrative of such combinations, conveying entertainment as well as information and ideas. The Ted Koppel "Nightline" show emphasizes key controversial issues, usually of a political nature.

The outline in Figure 6–9 is illustrative of a basic rundown sheet for a television interview program.

Figure 6–9 Interview rundown sheet.

VIDEO	AUDIO	
Open cold, studio interview set.	INTERVIEWER:	Welcome to "Interview," a weekly series of interviews with people in the news.
		Our guest today is Norris Politico, a former White House staff member who has become one of the world's largest and wealthiest international arms dealers.

continued

VIDEO		AUDIO
		This past week he made headlines by being accused of almost single-handedly starting a war between the southeast Asia nations of Maribia and Paquan by selling each arms after convincing each, separately, that the other was preparing an armed invasion.
First Commercial Break		
	Interviewer:	Before we get into the Maribia-Paquan affair, Mr. Politico, tell us how you first got into a White House position, and then into the arms trade.
	Politico:	Gives background info.
	Interviewer questions:	
	1. How does international arms selling work? Probe, if not answered by Politico: a. sources of funding (including Politico's) b. sources of arms (including Politico's) c. transfer of funds (Swiss bank accounts?; Politico accounts?)	
	2. Has Politico used his former White House contacts to obtain arms and customers? (Politico, in press, has said no.) Probe:	
Key in: Times article headline; Post article headline.	a. NY Times article, August 15, noting Politico used White House for military and foreign ministry liaison with other countries; b. Washington Post article, September 2, reporting that Politico had used Pentagon contacts to get priority arms purchases from U.S. defense contractors.	

Figure 6–9 *continued*

VIDEO	AUDIO
	3. How much money has Politico grossed and netted on arms deals during the past year?
	(In preinterview, Politico refused to say.)
	Probe:
Key in: copies of plane tickets	Network investigation turned up copies of air tickets showing he had made six trips to Switzerland in past two months, each trip for one day only. Repeat Swiss bank account query.
Second Commercial Break	4. Any moral compunction in selling arms?
	(In preinterview, Politico noted that a. somebody will do it, why not him? b. he sells only to countries whose policies he approves of; would never sell to an enemy of the U.S.)
	Probe:
	a. Maribia supports U.S. policies, but Paquan is considered to be unofficially aligned with some anti-American countries in Asia.
SOT: statement by Paquan foreign minister :15; statement by Maribia foreign minister :15.	b. The Paquan foreign minister recently stated that Paquan made contact with him in January, a full month before the Maribian foreign minister says he had initial contact with Politico.
	5. Rumor around Washington is that Politico has had not only the blessing but the assistance of the White House in his arms deals. Does Politico want to go on record with an outright denial? If so, why does he think the rumors are so persistent?
	Interviewer: Sums up interview.
	Politico: Final statement.
Commercial	Interviewer: Thank you, Norris Politico, for joining us on "Interview." Next week our guest will be ________. Good night.

The Discussion Program

Discussion programs are aimed toward an exchange of opinions and information and, to some degree, toward the arriving at solutions, actual or implied, of important questions or problems. They should not be confused with the interview, in which the purpose is to elicit, not to exchange.

The writer of the discussion program has to walk a thin line between too much and not enough preparation. It is not possible to write a complete script, partially because the participants can't know specifically in advance what their precise attitude or comment might be on any given issue or statement brought up in the discussion. On the other hand, a complete lack of preparation would likely result in a program in which the participants would ramble; it would present the moderator with the impossible task of getting everybody someplace without knowing where they were going. To achieve spontaneity, it is better to plan only an outline, indicating the general form and organization of the discussion. This is, of course, in addition to whatever standard opening, closing, and transitions are used in the program. This might include opening and closing statements for the moderator, introductions of the participants, and general summaries to be used by the moderator in various places throughout the program.

GAME SHOWS

A continuing staple of commercial television, game shows start early in the morning and fill program schedules through the afternoon and into prime time. The formats for the game show are many, ranging from the quiz of a contestant or an entire panel, to the performance of an audience member, to the participation of a celebrity. In all formats the participant is put into competition with another person or persons or with a time or question barrier. In each case someone is expected to solve some problem in order to achieve the specific goal of the game, that is, to win money or goods. The problem may be to stump a panel of experts, guess what someone else is thinking, answer extremely complicated or extremely elementary questions about some subject, or hit one's spouse in the face three out of five times with a custard pie.

Game shows are made to look spontaneous. Because the participants are in part or totally nonprofessionals in the acting field, it is not possible to write out a complete script. The readings would come out stilted. Yet, as with any well prepared program, as much of the continuity as possible is written out, and the routine or rundown sheets for many game shows are quite detailed. Opening and closing continuity, introductions, adlib jokes, questions, commentary introducing and ending sequences, and similar material are written out. The material should be flexible and adaptable to the spontaneity of the participants. The prepared material should be designed to fit the personality of the person hosting the show and should be developed in consultation with that host. In some audience participation shows virtually all of the dialogue, except that which will be given by the nonprofessional participants, is scripted. In many instances even the participants' supposedly ad-lib dialogue is prepared, even if only in outline form. Sequences, including routines, stunts, matches, or whatever the particular game show calls for, should be timed accurately beforehand so that the basic script fits into the required time length. Alternate material of different time lengths may be prepared in case the show begins to run too long or too short. The visual elements of the particular game should be stressed. Charts, pyramids, curtains, models, all bigger than life, are some of the set pieces for game shows.

MUSIC AND VARIETY SHOWS

Music

Until the advent of music videos in the 1980s, musical programs on television had all but disappeared. The work of the writer was principally in preparing intros and outros and occasional transition continuity. Music videos changed that, requiring highly imaginative story lines and/or creative uses of video effects to illustrate the theme or sound of the pop song.

Music video programs require little continuity to be successful; they are virtually all visual, the musical numbers themselves constituting the sound. Other musical presentations on television require more continuity, sometimes even explanations by the conductor (for example, a special with Leonard Bernstein conducting, or John Williams and the Boston Pops). Because the orchestra itself is stationary, imagination is necessary to make the visual picture continually interesting.

Variety

The term variety implies a combination of two or more elements of entertainment and art: a singer, a dancer, a stand-up comic, a comedy skit, a Shakespearean actor, a puppeteer, a ventriloquist, a pianist, a rock group. Depending on the personality who is the principal figure in the program, several of these elements would be incorporated in a manner that shows off the star to the best advantage. The basic variety show types are the vaudeville show, the music hall variety, the revue, the comic-dominated show, the personality program with guests, the musical comedy approach, and the solo performance.

The most important thing for the writer of the variety show to remember is that there must be a peg on which to hang a show. The writer must develop a clear, central theme, capable of being organized into a sound structure, with a unity that holds all the parts of the program together. Otherwise, each number will be isolated, and unless the audience knows what the next act is and especially wants to watch it, it would feel free at any time to tune in another station at the end of an act. The theme could be a distinct one or the continuity factor could simply be the personality of the star.

The variety show of early television merged into the personality-host variety program with guests representing different fields, principally those of the entertainment profession. The following excerpts from a routine sheet from a popular program of this genre in the 1980s, "The David Letterman Show" (Figure 6–10), illustrate the basic work of the writer. In addition, the writer may prepare personal continuity for each guest separately.

In some cases, as in Figure 6–10, the material prepared by the staff writer is minimal, merely providing transitions from one segment of the program to the next. Although it may seem overly simple to the serious writer, it nevertheless has to retain a consistency from show to show and reflect the approach of the program and its star.

EDUCATION AND INFORMATION PROGRAMS

Education and information programs cover many areas: formal instruction to the classroom, informal education to adults at home, technical updating to professionals, vocational preparation, industry training, and many others. Some of the programs are purely or principally instructional in nature. Others are primarily informational. Still others, with elements of education and information, are public relations oriented.

Corporate video is an especially fast-growing field, with many corporations and even small companies preparing informa-

OPENING ANNOUNCE | OPENING ANNOUNCE

FROM NEW YORK

IT'S LATE NIGHT WITH DAVID LETTERMAN. TONIGHT: TOM HANKS, BOARDWALK PITCH MAN ARNOLD MORRIS, ACTRESS LEA THOMPSON. ALSO: SUPERMARKET FINDS AND CURRENT EVENTS CHYRON QUIZ. AND NOW A MAN WHO. . . .

DAVID LETTERMAN!

* *

NEW YORK CITY VIDEO QUIZ #1 | ACT 1

THIS MAN GOT HIS JOB BECAUSE:

A) HE HAD THE REQUIRED SKILLS

B) HE DID WELL ON THE ENTRY EXAM

C) HE WAS TOO OLD TO CONTINUE IN MENUDO

[SERIES OF VISUAL QUIZ GAGS]

* *

TOM HANKS | ACT 2, 3, 4

TO USE A SHOW BUSINESS TERM, MY FIRST GUEST TONIGHT IS MONEY IN THE BANK. HE IS ALWAYS WELL PREPARED, EXTREMELY CHARMING AND WITTY WITHOUT POSING A THREAT. HE ALSO MAKES SOME PRETTY DECENT MOTION PICTURES, THE LATEST WITH JACKIE GLEASON CALLED "NOTHING IN COMMON." PLEASE SAY HELLO TO TOM HANKS.

* *

SUPERMARKET FINDS #1 | ACT 5

CAMPBELL'S CURLY NOODLE SOUP

-FROM CAMPBELL'S NEW LINE OF THREE STOOGE SOUPS. I LIKE THE SHEMP BROTH.

[SERIES OF SATIRES ON COMMERCIALS]

* *

LEA THOMPSON | ACT 6

MY NEXT GUEST HAS BEEN SEEN IN SUCH MOVIES AS "ALL THE RIGHT MOVES," "SPACE CAMP," "BACK TO THE FUTURE" AND THE CLASSIC "JAWS 3-D." SHE IS NOW STARRING IN THE MOVIE "HOWARD THE DUCK." PLEASE WELCOME ACTRESS LEA THOMPSON.

Figure 6–10 Script excerpts from "The David Letterman Show": opening announcement, visual quiz gag, guest introduction, satire on commercials, and second guest introduction. (Courtesy of National Broadcasting Company, Inc. © 1986, National Broadcasting Company, Inc. All rights reserved.)

tion, instruction, sales, and public relations TV programs, for internal and external use.

Formal Instruction

The writer of the formal education program is, above all, a planner. The writing of the program begins with the cooperative planning of the curriculum coordinator, the studio teacher, the classroom teacher, the educational administrator, the producer, the TV specialist, and the writer. The writer must accept from the educational experts the purposes and contents of each program. The writer should stand firm about the method of presentation; educators, by and large, are too prone to use television as an extension of the classroom, incorporating into the television program the outmoded techniques of teaching that dominate many classrooms. The most important thing to remember is to avoid the "talking head."

After determination of learning goals and contents, the length of individual programs and of the series is determined and the programs are outlined. The outline should carefully follow the lesson plan for each learning unit as developed by the educational experts. The important topics are stressed, the unimportant ones played down. The educational program does not have to be fully scripted, however. It may be a rundown or routine sheet, depending on the content and whether a studio teacher is used and to what degree that studio teacher is a professional performer.

Even in the outline stage a writer should explore the special qualities of television that can present the content more effectively than in the classroom, even when there is a competent classroom teacher. It is important to infuse creativity and entertainment into the learning materials. The TV lesson can use humor, drama, and suspense, and borrow liberally from the most effective aspects of entertainment programs. A writer should not be afraid of a liberal infusion of visuals. Good use of visual writing permits more concrete explanation of what usually is presented in the classroom. The classroom teacher frequently presents principles and explains with examples. Through television the examples can be infinitely more effective than the usual verbal descriptions. Television can show the real person, thing, place, or event itself.

The sequences in the script follow a logical order, usually beginning with a review and preparation for the day's material and concluding with introductory elements for follow-up in the classroom, including review, research, field projects, and individual study. Before beginning the script and before incorporating all the imaginative visual stimuli and the attention-getting experiences, one must know what is available in terms of the program's budget.

The word *motivation* is mentioned so often in education that we have become inured to it. Yet, it is still the key to TV watching and to learning. The better the "show" is, the better the student will learn. To make learning exciting, the material must be pleasurable and stimulating. Teaching is a form of persuasion. The instructional script should be developed as much as possible for a target audience. The educational experts involved are the best consultation source for determining how complex the concepts should be, what the students already know about the subject, and what the language level of the program should be. Although for different persuasive purposes, the instructional program may follow the organization of the commercial: get the students' attention, keep their interest, impart information, plant an idea, stimulate thinking about the subject, and, most important of all, motivate the students to create, through their own thinking, something new. The instructional program, like the good play, should increase in interest and intensity.

Corporate Video

The would-be writer for television will find himself or herself well advised to explore industry training, public relations, and sales divisions as likely sources for jobs. So-called corporate video grew rapidly in the 1980s, with many companies developing or expanding their audio-visual production areas to serve the above needs, sometimes within one division for the entire company, sometimes in separate divisions as noted above. Internal training programs serve entry level, mid-management, and top-level employees, and range from professional concerns (such as the development of new engineering processes or sales techniques for a new product) to staff concerns (such as uses of a new computer for office personnel or alternative skills training for maintenance people).

Corporate television scripts follow the same basic approaches as formal instruction scripts. Scripts do vary in terms of content, in terms of time and facilities requirements, and especially in terms of techniques required to convey some of the complicated and detailed industrial information.

Donald S. Schaal, as TV producer-director for Control Data Corporation Television Communications Services, said that "when you come to grips with scripting for industrial television, for the most part you might just as well throw all your preconceived ideas about creative/dramatic and technical writing in the circular file." Schaal said that attempts to transfer the classroom teacher to television have failed and that "unfamiliarity with what television could or could not do . . . resulted in a product which left just about everything to be desired. It lacked organization, continuity, a smooth succession of transitions and, in many cases, many of the pertinent details. . . . Since we think so-called 'training' tapes should *augment* classroom material and not supplant it, we soon realized that we could gain little but could lose everything by merely turning an instructor loose in front of the tube to do exactly what he does in person in the classroom. . . . The videotape he needs for his classroom *must* provide something he cannot conveniently offer his students in person."

Roger Sullivan, as director of education for the Commercial Union Assurance Companies, stated that it is important for business video writers to understand the objectives of the particular video being produced. There are two general objectives: management wants the employees to learn skills that enable them to carry out their jobs more effectively and employees want personal growth and the ability to do a better job that is recognized by management. Within each program there are a series of modules of specific objectives that the writer must break down and accomplish with the most effective visual presentation for each module.

Corporate video writer-producer-director Ralph DeJong stresses the need for entertaining story-telling techniques, but cautions that the corporate TV production "demands that the writer be acutely aware of the relationships among people, processes, equipment, and institutional philosophies and goals." He adds that "the scriptwriter must identify the target audience and develop a profile of its interests and familiarity with the subject matter."

THE PLAY

Two types of drama have emerged in television that present different approaches for the playwright. Where the writer is creating a play from scratch, that is, a drama with new characters, plot line, and setting, the basic principles of creative playwriting hold. New TV series, drama specials, and so-called miniseries fall into this category. Where the writer is preparing an episode for an established series, the characters, dialogue, setting, and types of plot lines are already established, and the writer must combine creativity with strict adherence to the formula already set for the program.

Principles

The rules of playwriting are universal. They apply in general to the structure of the play written for the stage, film, or television. The rules are modified in their specific applications by the special requirements of the particular medium. One should not assume that because there are rules of playwriting that the art can be taught. Genius and inspiration cannot be taught, and playwriting is an art on a plane of creativity far above the mechanical facets of some of the phases of continuity writing.

Sources

The writer may find the motivating ingredient for the play in an event or happening, in a theme, in a character or characters, or in a background. The writer may initiate the preliminary thinking about the play from an idea. Another source for the play may be the social environment. The character may be several people rolled into one. The sources of the play usually are only germs of ideas. The writer should write, ideally, out of personal experiences or knowledge so that the play may have a valid foundation. However, if the writer is too close, either emotionally or in terms of time, to the life-ingredients of the play, it will be difficult to heighten, condense and dramatize.

Play Structure

Modern drama has emphasized character as most important, although many TV series sacrifice character for plot action. Ideally, the actions that determine the plot are those the characters must take because of their particular personalities and psychological motivations. The dialogue is that which the characters must speak for the same reasons. The three major elements in the play structure—character, plot, and dialogue—all must be coordinated into a consistent and clear theme. This coordination of all elements toward a common end results in the unity of the piece, a unity of impression. The characters' actions and the events are not arbitrary, and the audience must be prepared for the occurrence of these actions and events in a logical and valid manner. *Preparation* is the term given to the material that thus prepares the audience. The background and situation also must be presented; this is the *exposition*. Another element the playwright must consider is the *setting*, which the playright describes in order to create a valid physical background and environment for the characters.

Only after one understands and can be objective about the characters, theme, situation and background, can one begin to create each of them in depth. A playwright should do as much research as possible to become acquainted fully with the potentials of the play.

After the characters have been created, the playwright is ready to create the situation, or plot line. This should be done in skeletonized form. First, a play needs a conflict. The *conflict* is between the protagonist of the play and some other character or force. A conflict may be between two individuals, an individual and a group, between two groups, between an individual or individuals and nature, between an individual or individuals and some unknown force, or between an individual and the inner self. The nature of the conflict will be determined largely by the kinds of characters involved. After the conflict has been decided upon, the plot moves inexorably toward a *climax*, the point at which one of the forces in conflict wins over the other. The play reaches the climax through a series of complications. Each *complication* is, in itself, a small conflict and climax. Each succeeding complication literally complicates the situation to a greater and greater degree until the final complication makes it impossible for the struggle to be heightened any longer. Something has to give. The climax must occur.

The complications are not arbitrary. The characters themselves determine the events and the complications because the actions they take are those, and only those, they must take because of their particular motivations and personalities.

Concepts of Playwriting

Unity

One of the essentials that applies to all plays, regardless of type or style of production, is the unity of action or impression. There should be no elements within the play that do not relate in thorough and consistent fashion to all the other elements, moving toward a realization of the purpose of the playwright.

Plot

The plot structure of a play is based on a complication arising out of the individual's or group's relationships to some other force. This is the conflict, the point when the two or more forces come into opposition. The conflict must be presented as soon as possible in the play, for the rest of the play structure follows and is built upon this element. Next comes a series of complications or crises, each one creating further difficulty in relation to the major conflict, and each building in a rising crescendo so that the entire play moves toward a final crisis or climax. The climax occurs at the instant the conflicting forces meet head on and a change occurs to or in at least one of them. This is the turning point. One force wins and the other loses. The play may end at this moment. There may, however, be a final clarification of what happens, as a result of the climax, to the characters or forces involved. This remaining plot structure is called the *resolution*.

Character

Character, plot, and dialogue comprise the three primary ingredients of the play. All must be completely and consistently integrated. In modern dramaturgical theory character is the prime mover of the action, and determines plot and dialogue. The character does not conform to a plot structure. The qualities of the character determine the action. The character must be revealed through the action; that is, through what the character does and says, and not through arbitrary description or exposition. Character is delineated most effectively by what the individual does at moments of crisis. This does not imply physical action alone, but includes the concept of inner or psychological action.

Dialogue

Inasmuch as the play does not duplicate real people or the exact action of real life, but heightens and condenses these elements, the dialogue also has to be heightened and condensed rather than duplicated. The dialogue must truly conform to the personality of the character speaking it, must be completely consistent with the character and with itself throughout the play, and must forward the situation, the showing of the character and the movement of the plot.

Exposition

Exposition, the revelation of the background of the characters and situation and the clarification of the present circumstances, must not be obvious or come through some device, such as the telephone conversation, the servant, or the next-door neighbor. It must come out as the action carries the play forward and must be a natural part of the action. The exposition should be presented as early as possible in the play.

Preparation

Preparation, too, must be made subtly. Preparation, or foreshadowing, is the unobtrusive planting, through action or dialogue, of material that prepares the audience for subsequent events, making their occurrence logical and not arbitrary.

Setting

Setting is determined by the form of the play and the physical and mechanical needs of the play structure. Setting serves as locale, background, and environment for the characters of the play.

Television's Special Characteristics

The special characteristics of the television audience require a special approach on the part of the playwright. You may combine the subjective relationship of the viewer to the television screen with the electronic potentials of the medium to create a purposeful direction of the audience's attention. You may direct the audience toward the impact of the critical events in the character's life and toward the subjective manifestations of the character's existence. The ability to focus the viewer's sight, attention, and even feelings specifically permits the writer to orient the consciousness of the audience closely to the inner character of the person on the screen.

The hour drama is really only about forty-two minutes long, the half-hour drama twenty-one minutes in length. Even the two-hour dramatic program permits only about eighty-two minutes for the play. The television play should be extremely tight; it should have no irrelevancies. It should have as few characters as possible and one main, simplified plot line, containing only material relating to the conflict of the major character or characters.

Dramaturgical Concepts for Television

Unity

The most important changes in the unities as applied to television relate to time and place. Television can transcend boundaries of time and place that even the most fluid stage presentation cannot match.

Plot

The dramaturgical rules relating to plot apply to the television play as to the stage play. The problem of time, however, necessitates a much tighter plot line in the television play, and a condensation of the movement from sequence to sequence. The short time for the television play requires the plot to be the essence of reality, to contain only the heightened extremities of life. Aim for the short, terse scene. In recent years the success of this approach has been exemplified by "Hill Street Blues" and the many series that copied its style.

Although the emphasis on plot, because of the time factor, seems to make this the motivating factor in the television play, the exploration of television's intimacy and subjectivity potentials enables the writer to delve into character and to use it as a plot-motivating element.

Character

A playwright should not use unneeded people. A character who does not contribute to the main conflict and to the unified plot line does not belong in the play. Only a character who is essential should be put in the script.

Dialogue

Television requires one significant modification in the use of dialogue to forward the situation and to provide exposition. The visual element can often substitute for the aural. If a playwright can show the situation or present the expository information through action instead of through dialogue, it should be done. Location shooting and the close-up have made it possible to eliminate time-consuming dialogue in which the character describes things or places. The playwright can concentrate not only on action but on "reaction," keeping the dialogue at a minimum and the picture the primary object of attention.

Exposition

The short time allotted to the television play permits only a minimum of exposition. Because the conflict should be presented almost as soon as the television drama begins, the exposition must be highly condensed and presented with all possible speed.

Preparation

The writer should prepare the audience in a subtle and gradual manner for the subsequent actions of the characters and the events of the play. Nothing should come as a complete surprise.

Setting

Television drama essentially conforms to the play of selective realism in content and purpose, and realistic settings usually are required.

The Scenario and Outline

The scenario, the treatment, the outline, and the summary give the producer and/or script editor a narrative idea of what the play is about: the plot line, the characters, the setting and maybe even bits of dialogue. Most producers/editors can tell from this narrative whether the play fits the needs of the particular program. The scenario and treatment are usually longer than the outline and summary, perhaps as much as a fifth of the entire script. The summary and the outline may be only two or three pages, in effect providing a preliminary judgment prior to a preliminary judgment. Some producers/editors want to see a summary or outline first, then a scenario or treatment and, finally, the complete script.

The Manuscript

The television manuscript should have all the characters clearly designated, the dialogue, the stage directions, the video and audio directions and, in the Hollywood-style filmed play, the shot designations. The most frequently used form is the two-column approach: the right-hand column containing all of the audio—that is, the dialogue plus the character's movements—and the left-hand column containing the video—that is, the mechanical and electronic effects.

Another manuscript approach is to place all of the material, video and audio, together, right down the center, similar to the stage play form, or solely in a left-hand column or right-hand column, leaving the other side free for the director's notes. The names of the characters should be typed in capital letters in the center of the column, with the dialogue immediately below. Video and audio directions are usually differentiated from the dialogue by being in parentheses and/or in capital letters and/or underlined. Script editors prefer that dialogue be double-spaced, with double-spacing between speeches.

The filmed play does not have the continuous action that still marks many taped plays. The filmed play has a break at each cut or transition. That is, each sequence may last two seconds to two minutes or longer. Between sequences the director can change sets, costumes, makeup, reset lights and cameras, and even reorient the performers. The filmed play is not shot in chronological order, as the taped play usually is. All the sequences taking place on a particular set or at a particular locale, no matter where they appear in the script, are shot over a contiguous period of time. Then the entire cast and crew move to the next set or locale and do the same thing. The writer writes shots, not scenes. Each shot is set in terms of a picture rather than in terms of character action, although the latter should, in all plays—filmed or taped or live—be the motivating factor. The writer states the place, such as INTERIOR or EXTERIOR, and the shot, such as FULL SHOT or CLOSE-UP. The writer also describes the setting, states the characters' physical relationships to the set and their proximity to each other, and then presents the dialogue for that shot. The individual shots are numbered in consecutive order so that the director may easily pick out any sequence(s) desired for initial shooting, retakes, or editing. (See Figures 6–1 and 6–2 for examples of the two formats.)

One program using the single-column format, with scenes arranged so that it can be shot on tape with live-type continuity, or on film, out of sequence, is "The Cosby Show." The first few pages of an episode entitled "Vanessa's Bad Grade," written by Ross Brown, are presented in Figure 6–11.

Figure 6–11 One television play format: "The Cosby Show" sitcom. (Written by Ross Brown. Courtesy of "The Cosby Show.")

THE COSBY SHOW

"Vanessa's Bad Grade"

SHOW #0212-13

CAST

Cliff Huxtable .... Bill Cosby

Clair Huxtable .... Phylicia Ayers-Allen

Denise Huxtable .... Lisa Bonet

Theo Huxtable .... Malcolm-Jamal Warner

Vanessa Huxtable.. Tempestt Bledsoe

Rudy Huxtable .... Keshia Knight Pulliam

Robert.. Dondre Whitfield

Announcer (V.O.) .... TBA

SET

ACT ONE		PAGE
Scene 1:	INT. KITCHEN—MORNING (DAY 1)	(1)
Scene 2:	INT. LIVING ROOM—THAT AFTERNOON (DAY 1)	(7)
Scene 3:	INT. KITCHEN—CONTINUOUS ACTION (DAY 1)	(9)
Scene 4:	INT. LIVING ROOM—TWO DAYS LATER—AFTERNOON (DAY 2)	(15)
Scene 5:	INT. RUDY & VANESSA'S ROOM—CONTINUOUS ACTION (DAY 2)	(19)
ACT TWO		
Scene 1:	INT. RUDY & VANESSA'S ROOM/HALLWAY—THAT NIGHT (DAY 2)	(21)
Scene 2:	INT. LIVING ROOM—LATER THAT NIGHT (DAY 2)	(23)
Scene 3:	INT. HALLWAY/RUDY & VANESSA'S ROOM—CONTINUOUS ACTION (DAY 2)	(26)
Scene 4:	INT. LIVING ROOM—CONTINUOUS ACTION (DAY 2)	(28)

continued

SET *continued*

Scene 5: INT. DENISE'S ROOM—CONTINUOUS ACTION (DAY 2) (32)

Scene 6: INT. HALLWAY—CONTINUOUS ACTION (DAY 2) (34)

Scene 7: INT. DENISE'S ROOM—CONTINUOUS ACTION (DAY 2) (35)

Scene 8: INT. RUDY & VANESSA'S ROOM—CONTINUOUS ACTION (DAY 2) (41)

Scene 9: INT. KITCHEN—CONTINUOUS ACTION (DAY 2) (44)

Scene 10: INT. DENISE'S ROOM—A LITTLE LATER THAT NIGHT (DAY 2) (50)

ACT ONE

Scene 1

FADE IN:

INT. KITCHEN—MORNING (DAY 1)
(Cliff, Clair, Denise, Vanessa, Rudy)

(RUDY SITS AT THE TABLE BLOWING BUBBLES IN HER GLASS OF MILK. CLAIR ENTERS)

CLAIR

Rudy, don't blow bubbles with your straw.

RUDY

Okay.

(RUDY SUCKS MILK INTO THE STRAW, PUTS IT IN HER BOWL)

CLAIR

Rudy, put the straw down and drink your milk.

(RUDY DRINKS THE MILK OUT OF THE BOWL)

CLAIR (CONT'D)

All right, that's enough.
Breakfast is over. Go brush your teeth and get ready for school.

(CLIFF ENTERS)

CLIFF

Hey, Pud.

RUDY

Hi, Daddy. Don't play with the straw.

(RUDY EXITS)

Figure 6–11 *continued*

<u>ACT ONE</u> *continued*

CLAIR

How are you feeling?

CLIFF

Hmmm.

CLAIR

When did you get in?

CLIFF

Hmmm.

CLAIR

Poor baby. That's the third time this week.

CLIFF

Clair, I would say that during my career, I've delivered about three thousand babies. Somehow, almost all of them decided to be born between two and five a.m. on the coldest winter nights of the year. There must be an all-weather radio station just for babies. When they hear, 'It's two a.m. Heavy snows and arctic winds,' they shoot for daylight.

CLAIR

Maybe we should forget about the movie tonight.

CLIFF

No, no. We're going.

CLAIR

You're too tired.

CLIFF

I'll be fine.

CLAIR

That's what you always say. Then as soon as the lights go out, so do you. Why don't we go Friday night?

(CLIFF PICKS UP THE NEWSPAPER)

continued

ACT ONE *continued*

CLIFF

Listen to this, Clair. 'Friday's forecast: clear and warmer.' No babies.

CLAIR

Then Friday it is.

(VANESSA ENTERS)

VANESSA

Mom?

CLAIR

Yes?

VANESSA

Can we go shopping? I need a new sweater.

CLAIR

Vanessa, you just got some new sweaters for Christmas.

VANESSA

I know, but Robert's seen me in all those.

CLIFF

You could wear them inside out.

VANESSA

But Robert's taking me to the school dance on Friday and I really want to wear a new sweater. I've even got one picked out.

CLAIR

Oh?

VANESSA

I want one exactly like the one Denise got.

CLAIR

Why don't you ask Denise if you can borrow hers?

VANESSA

Mom, she's not going to let me have her sweater.

Figure 6–11 *continued*

<u>ACT ONE</u> *continued*

CLAIR

She might. Why don't you tell her why you want it and ask her nicely? You may be surprised.

VANESSA

All right. I'll try. Denise . . .

(SHE EXITS)

CLIFF

You want to take any bets on this one?

CLAIR

It could happen. Denise has been in that position with Sondra, so she might be understanding.

CLIFF

Okay. A jumbo box of popcorn at the movie says Vanessa doesn't get the sweater.

CLAIR

I'll take that bet.

DENISE (O.S.)

Ha! Are you kidding? No way. You're not getting it.

(DENISE AND VANESSA ENTER)

VANESSA

But, Denise, I asked nicely.

DENISE

I don't care.

VANESSA

But it's for the dance. Robert is taking me.

DENISE

I haven't even worn that sweater yet. Bye, Dad.

VANESSA

Why don't you wear it today, and then you've worn it. Bye, Dad.

continued

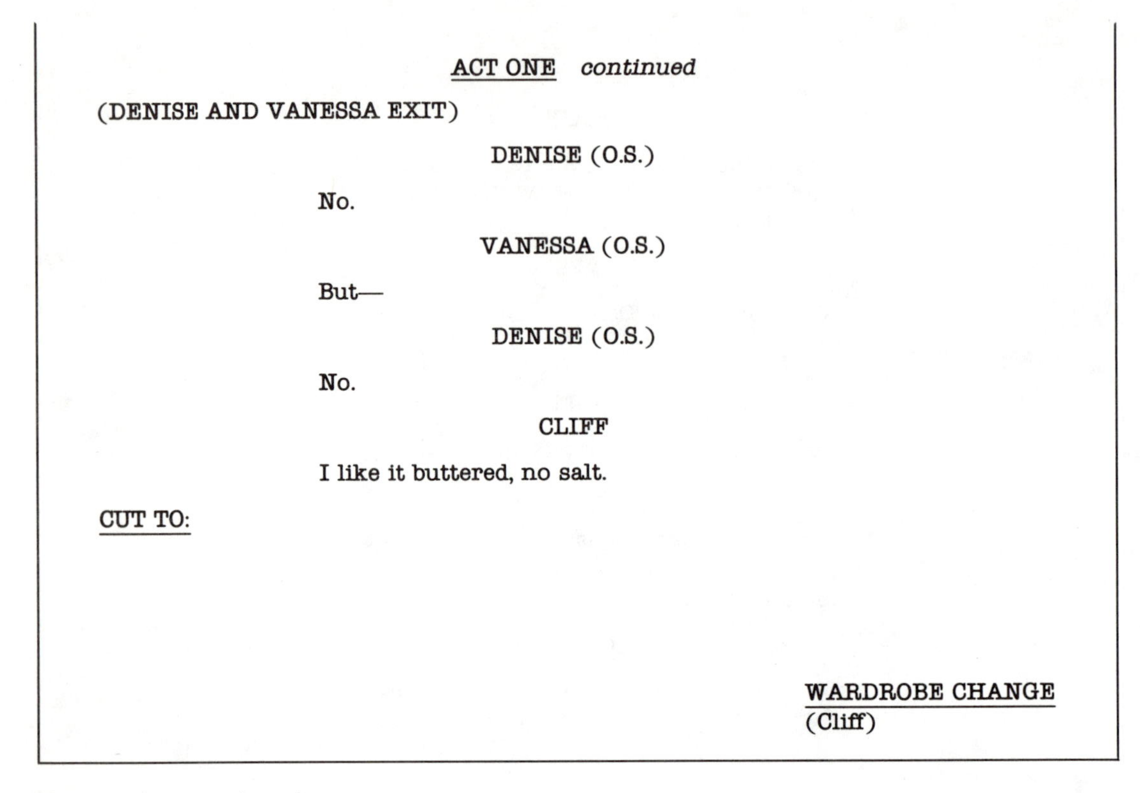
ACT ONE *continued*

(DENISE AND VANESSA EXIT)

DENISE (O.S.)

No.

VANESSA (O.S.)

But—

DENISE (O.S.)

No.

CLIFF

I like it buttered, no salt.

CUT TO:

WARDROBE CHANGE
(Cliff)

Figure 6–11 *continued*

Rod Serling, who was one of television's most articulate as well as prolific writers, called TV a medium of compromise for the writer. He was concerned that the writer could not touch certain themes or use certain language. He criticized television because of "its fear of taking on major issues in realistic terms. Drama on television must walk tiptoe and in agony lest it offend some cereal buyer. . . ." Despite these restrictions, he felt that "you can write pretty meaningful, pretty adult, pretty incisive pieces of drama."

As a writer one is, with relatively few exceptions, dependent on the commercial mass media for one's existence; yet, despite the restrictions put upon a writer by sponsors, networks, and production executives, the play is still the prime mover, the one element upon which all other elements of the production must stand or fall. With a script of high quality, with writing of ethical and artistic merit, a writer may at least take pride in having made a significant effort to fulfill some of the mass media's infinite potentials.

BIBLIOGRAPHY

Armer, Alan A. *Writing the Screenplay*. Belmont, CA: Wadsworth, 1988.

Book, Albert C., Norman D. Cary, and Stanley I. Tannenbaum. *The Radio and Television Commercial*. Chicago: Crain Books, 1984.

Bliss, Edward J., Jr. and John M. Patterson. *Writing News for Broadcast*. New York: Columbia University Press, 1978.

Blum, Richard A. *Television Writing: From Concept to Contract*. Boston: Focal Press, 1984.

Garvey, Daniel E. and William L. Rivers.

Newswriting for the Electronic Media. Belmont, CA: Wadsworth, 1982.

Hagerman, William. *Broadcast Advertising Copywriting*. Boston: Focal Press, 1989.

Hilliard, Robert L. *Writing for Television and Radio*. 5th ed. Belmont, CA: Wadsworth, 1989.

Meeske, Milan D. and R. C. Morris. *Copywriting for the Electronic Media*. Belmont, CA: Wadsworth, 1987.

Newsom, Doug and James A. Wollert. *Media Writing: Preparing Information for the Mass Media*. 2d ed. Belmont, CA: Wadsworth, 1988.

Orlik, Peter B. *Broadcast Copywriting*. 2d ed. Newton, MA: Allyn & Bacon, 1982.

Strunk, William, Jr. and E. B. White. *The Elements of Style*. 3d ed. New York: Macmillan, 1979.

7

Directing

Jeffrey Lukowsky

The first television show I directed was a two-hour Off-Broadway musical staged in an experimental theater space. It was very complicated: singers, dancers, props, even a seventeen-piece band on stage, and all of it had to be captured by three television cameras and made into one pleasing, coherent television program. As television director, I was sitting off-stage in the TV control booth (control booths will be discussed later in this chapter). As the zero hour approached, less than ten seconds to showtime, I froze. I thought I was having a heart attack. I could not speak and I could not move. I was certain that my career in television was over before it had really begun.

Suddenly, as the clock struck zero, I was issuing commands as fast as I could speak. When my assistant informed me that we had just passed the one hour mark of the show, I was struck dumb. I felt as if I had only been at work for five minutes. Yet I remember looking at the clock at the one hour and fifteen minute mark and feeling as if I had been at work for two days. Such is the excitement and intensity of television direction, that subjective, or personal, time often feels as if it has little or no relation to objective, or clock, time.

Two months later, after the program was well received by the public, I paused to reflect upon this first experience directing a large-scale television production. While directing, I did not feel in control of all the people and machines as if I were outside them or above them. (The usual term is *above* because we conceptualize hierarchical authority on the vertical axis.) I thought of myself as in the center of this magical electronic universe. All the various messages coming into me had to be patterned or rearranged, and then sent back somewhere (as it turns out, to the master video recording deck). My humble function, I thought, was to be the big switching mechanism of all the information, and there was a lot of it. I watched six television monitors at once, listened to four audio lines, pushed buttons, issued commands, and somehow got the entire process to come out as one satisfying television piece. The *somehow* of this process is exactly what this chapter is about. What is a television director and how does the director produce that final program from so many initially separate elements?

THE MIND OF A DIRECTOR

Two principal aspects of a director's thinking are *technical* reasoning and *aesthetic* reasoning. An analogy is the experience of taking still photographs, even with an instamatic camera.

Many of the pictures we took with our first still cameras disappointed us. They were out of focus, too dark or too light, the composition was poor, or Uncle Willy's head was missing from his body. So our education began first through the recognition that what we saw in our minds (or through the lens) was not what we got. Then came the second recognition: the only way to get good pictures was by learning all sorts of technical principles and, at the same time, by continuing to practice with the camera. We had to learn about f-stops, shutter speeds, light meters, film stocks, indoor lighting, and so on. But even after most of the pictures were in focus and properly developed, we still needed excellent composition and interesting lighting. This is where the art begins and mere technique leaves off. Even users of automatic cameras who need to do nothing but "point and push the button" know that many of their photographs are dissatisfying because of poor composition, or the action was caught too late. So getting good photographs was not simply the mastery of technique.

Once we acquired varied technical competencies, we began to relax with the equipment and feel confident as photographers. Technical considerations were no longer a burden but a creative challenge. At that point we first became free to create aesthetic photographs. Only by emancipating ourselves from the worries of technical constraints are we able to "see through" the equipment and make art. Learning how to use the equipment was initially an end in itself. But once we did that, we could relax and remember that the equipment was a means to something and not an end in and of itself. The end, of course, was to make photographs that pleased us and pleased others.

To sum up this first part of our analogy, aesthetic visualizations are only possible after placing the technical dimensions of equipment at the service of one's creative desires. Of course, artistic photographs result not only from each artist's creative thoughts, but also from the creative use of various technical dimensions in relation to each other. But a professional photographer is able to move back and forth between an emphasis on technical considerations and on aesthetic considerations. Many professionals talk of making technical decisions almost automatically or, as it were, below the level of direct self-awareness. This comes only after much learning and practice. The purpose of this learning and practice is to allow oneself to focus on the most important goal of visual work: to make artistic imagery, and not merely technically proficient imagery.

The other part of our analogy deals with where the television director fits. The television director not only must think like the still photographer, but must think about two other complicated phenomena: motion and sound. Let us look at the introduction of motion into the picture and, keeping it simple, think of the television director, alone, with one television camera shooting all the action for a live community cable show. Whatever the director does with the camera, we are certain to see it.

As the show begins, we, the audience, see two people discussing a current and controversial social issue. The director is standing in front of the two guests with the television camera. The camera is, so to speak, our eyes. If it is stage front, then the camera is in the position of the audience and we see the two guests facing us as if we were in a traditional theatre performance, except, of course, that the image is being broadcast to our homes. (Much early cinema was a visualization of traditional proscenium stage plays. It took time for an independent cinema aesthetic to develop.) The camera does not move for ten minutes. For twenty minutes, then thirty minutes it has not moved, then it does not move for one hour! Would the audience still be awake? Very few people can even look at a great photograph for twenty minutes without leaving it and later returning, and when we look at a motion medium we expect some dynamic movement.

A director knows that the audience wants a moving picture; therefore, the director moves the camera to create television motion. Much of this movement will appear graceful and have some motivation behind it. By *motivation*, we mean that the camera movement is not gratuitous but is for some purpose (e.g., a close-up on one of the speakers during a particularly heated moment in the discussion). But in our one-camera shoot the audience might see a great deal of motion irrelevant to the main point of the show. Let's say one speaker brings in a large architectural model to be referred to at various points in the program. Every time that speaker needs to refer to the model, our director has to move the camera to the model, then back to the speaker, and so on. This disrupts the flow of the program with the introduction of *waiting time*. Clearly, we have a problem here. Camera movement can be used as a pleasing aesthetic technique; indeed, it is one of the most basic television and film techniques. But not all camera motion will be pleasing because we may want to get from one visual emphasis to another without having the audience perceive all the movements in between those two emphases. One of the great joys of motion media is that long stories can be told in short periods of time through the use of space-time compression. By the time broadcast television became widespread in the U.S., film audiences had become accustomed to the dynamic pace of motion pictures through editing techniques. Broadcast television, to become commercially successful, had to provide the audience with, among other things, a dynamic motion medium. How was this possible?

The film director provides us with motion within scenes by moving cameras and talent around a set. But the film director also provides a different order of motion through the juxtaposition of sequences of images—more commonly known as film editing. Once the film is developed, the director can rearrange film segments into a final order. But what can our television director do with the one camera and two talk show guests?

One answer is to have the television director edit the television show. But we just said that the show was live; the television camera is electronically imaging the talk show and instantaneously transmitting it to the public as one continuous motion image. The answer is a live editing device! And this device, or *black box*, is a *camera switcher*. This enables the director to instantly "switch," or choose from more than one camera. For example, with three cameras, one could be focused on the architectural model to which our guests keep pointing, and the other two could focus on the guests (in various ways, to be discussed under the topic of camera placement). All three cameras are connected to the switcher and only one camera can be selected at any one time (by pushing buttons reading *Cam. #1*, *Cam. #2*, and *Cam. #3*). How does our director proceed?

The first change is that the director can no longer be a one-person operation. The director needs three camera operators to position the cameras in order to have a range of directorial choices. Because the director cannot look through three cameras at once, there are three television monitors to look at—each monitor representing one camera. Because the show is live, the director must always have one camera button on the camera switcher selected or the audience will not see any image. Suppose the TV audience sees both talk show guests speaking to one another. The director, located at the television studio, knows that we are seeing that particular two-person shot because the director sees it on the camera 2 monitor and has pushed the Cam. #2 button on the switcher. The switcher sends the image through the rest of the television broadcast system and it is received at home.

While the audience is watching the two-person shot, the director is looking at camera monitors 1 and 3 to see what shot should follow the current one. If the director does not like any of the images, a command can be issued to one or both camera operators, usually via headsets, to alter their camera positions. The director is then able to select the next shot by quickly pushing a button on the camera switcher. The audience reads the change as it would a film edit—a cut from one shot to the next. If the juxtaposition of shots (at the cut or switch point) is aesthetic, then the images seem to possess a natural flow. The change in camera positions does not call attention to itself and the audience can focus on the content of the discussion. If the switch from one camera to the next is abrupt or illogical, then the audience loses concentration.

As discussed, multicamera environments entail more than just one camera. Needed are a switcher, camera monitors, headsets, and a lot more equipment to be mentioned later in this chapter. All the equipment except for camera, lights, and microphones are removed from the set or stage and put into a control booth. Even so, it takes special on-stage camera and sound equipment to ensure that the audience hears the guests talking and not the sound of equipment being moved.

In the mid-1960s videotape recorders became commercially viable, and broadcast television entered a new era. Television could now be recorded for use at a later date. Indeed, segments of recorded tape could be electronically edited in desired sequence. The one-camera director could go back to being a one-camera person again, preferably not on live television. The use of one television camera recording images on a videotape recorder for use at a later date is quite naturally called shooting video cinematic style. The implication is that the director will spend time editing recorded images into a final, desired sequence.

The point of the analogy of the television director to the still photographer is to highlight the relationship between technical reasoning and aesthetic reasoning. Technical knowledge is required to produce competent photographs. But competency is only a prelude to artistry. The television director needs many technical skills to achieve the freedom that permits exploration of the aesthetics of the medium.

The television director watches many television monitors at once, gives commands to camera operators, listens to many audio sources, watches the clock, cuts to commercials, has titles appear on screen, gives commands to videotape machine operators, and so on. And all of this must become second nature.

As with the still photographer, if the television director is burdened by technical constraints and rules, he or she will never use the medium to produce artistic imagery. It takes many years of hard work to master all the varied aspects of broadcast television production. Then the television director needs to develop yet another skill, that of astutely motivating and directing production teams and talent through many hours of difficult work. Directors need to be proficient at handling human relationships as well as technical-aesthetic relations. The best directors are true authorities; people feel confidence in all aspects of their leadership, including the belief that the director cares for each of them individually. Directors who do not exhibit these human communication strengths rarely produce first-rate work.

CREATING TELEVISION IMAGES

Professional television directors know every major aspect of television production. Successful directors never stop learning. New tools and new imagery are constantly developed and tested. The smart directors are always in touch with the broadcast television industry as a whole (and they are usually well versed in the film industry, too). The novice television director has to begin his or her training at some point in this complex process. No single task or single set of skills can be termed the correct point at which to embark on the road to becoming a successful television director; television directors have great variety in their backgrounds and methods of training.

Some directors' initial training in the television industry may have been scriptwriting. Others may have begun as camera operators, still others as control room technicians. Most would agree that the successful television director must have a fine sense of connecting separate moving visual images in order to create meaningful and aesthetic experiences for the audience. The traditional method for teaching these skills begins with an analysis of visual composition followed by shot selections that exemplify preferred compositions. From here, one moves into a study of three basic types of motion: motion within a camera frame, camera motion, and motion among frames (switching and/or editing). These types of motion are related; for example, actors are often moving within the studio setting in relation to camera movement. This related movement of television cameras and talent is often a source of more dynamic television images than those created by the movement of just one or the other.

Camera Shots

One of the most important decisions a director makes is *shot selection*, or plotting shots. Television cameras can provide us with many points of view, and the director must decide how each event will be shown. Shot selections are a result of many considerations: the content of each particular scene, emphasis on desired meaning, aesthetic considerations (good composition), dramatic impact, maintenance of viewer interest, etc. The director uses a commonly understood vocabulary for shot selection, which is relayed to the camera operators. In television, the most basic shot vocabulary is based upon a single human body. Below are the commonly used shot terms (see Figure 7–1).

EXCU (sometimes *XCU*—extreme close-up, shows only a certain portion of the face as indicated by the director.

CU—close-up, a shot of the head only. Some directors call this a big, or tight, CU.

MCU—medium close-up, from the shoulders up. Also called a shoulder shot.

MS—medium shot, usually from the waist up.

LS—a full-length shot of the person.

One has to be careful with the use of these abbreviated terms found in the visual column of the script. A CU to one director may mean from the shoulders up, while to another director it means a face shot. To avoid confusion, some directors use terms that refer to portions of the human body, e.g., knee shot (from the knees up), waist shot (from the waist up), etc. These terms are a bit too cumbersome for many script templates which are produced on the standard eight and one-half by eleven inch sheets. Therefore, many directors will spell out what they mean by their abbreviations.

Sometimes, a combination shot is used. One of the most famous is the close-up long shot. In this situation, we have a close-up of one person in the foreground, and a long shot of another subject in the background. With current lens technology you can just about keep both subjects in perfect focus. This combination shot is frequently used in dramatic situations involving a dialogue between two characters.

With the development of the *zoom* lens, it became possible to give the effect of movement towards or away from the subject during a single continuous shot. The zoom lens contains multiple, moving elements that respond to a simple maneuver by the camera operator. As the camera operator moves the lens control, the lens progressively changes from a wide shot to a close-up or vice versa. In other words, a zoom lens is a combination of many fixed focal lenses. It allows directors to work with lighter equipment since cameras no longer need to have turrets with two or more lens mounts. Under the older equipment systems, a director had to stop

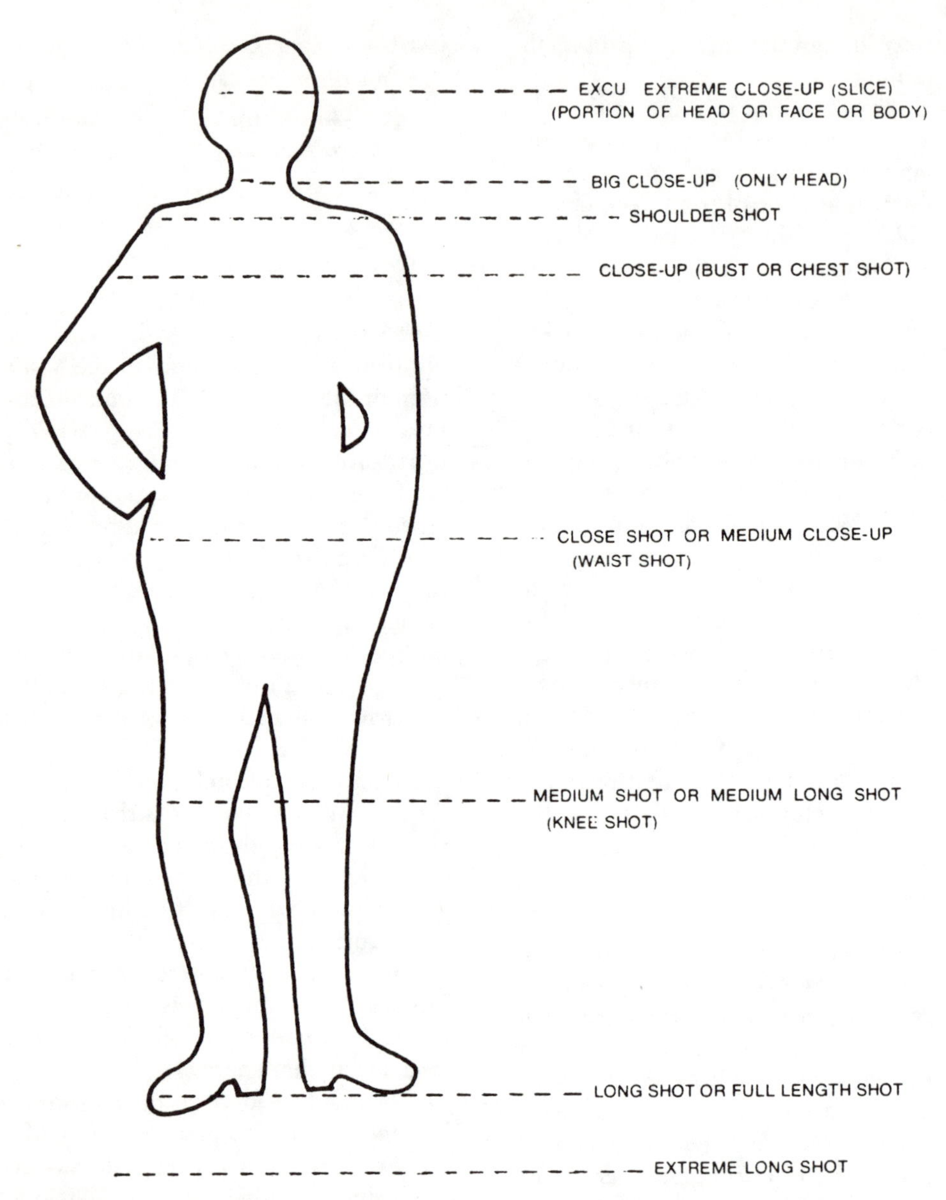

Figure 7–1 Commonly used camera shot terms. (Material prepared by Tom C. Battin.)

and move a new lens into position for every shot change. Now a director moves a control and the new shot is ready to go in two seconds. In addition, the zoom lens allows the audience to watch the zoom movement actually take place.

Zoom lenses were quite controversial when they first were widely used. It is important to know the difference between a zoom lens magnification and the use of a *fixed focal length lens* (or *prime lens*) which can only get us closer to the subject by moving the camera closer to the subject. Zoom lenses move the subject, or the scene, closer or farther away from the viewer through a sophisticated process of magnification. The three spatial planes, foreground, middleground, and background, maintain the same relationship to each other when one zooms in or out. In short, the entire image is uniformly moved towards or away from the viewer.

Fixed focal length lenses have no adjustable magnification. They work within limited ranges depending on their focal length. To bring us closer to the subject requires camera movement towards the subject (known as a *dolly-in*). This gives the viewer a sense of entering the scene, of being brought into the scene, rather than having the scene brought to them—as is the case in lens zooming. When one moves the camera towards or away from the subject the three spatial planes change their relations to each other. It is this fact that gives us the sense of being more within the scene than we feel when the scene is brought to us via the zoom lens.

The astute director wisely controls the use of zoom lens shots. Properly used, zoom lenses produce exciting and dramatic shots. They are versatile, easy to use, and make camera operations less cumbersome. Nonetheless, zoom shots simply cannot replace camera movement in terms of providing the viewer with a greater sense of being in the scene. In short, the zoom lens should not be overused.

Camera Movement

Camera shots and camera movement are two of the most basic visual tools that the television director must successfully control. Selection of appropriate camera shots requires knowledge of major principles of visual composition, which will be discussed later in this chapter. Camera movement usually alters the composition of the original shot choice. Television directors not only must know the classical rules of pictorial composition but they must also couple these rules to the motion variable. These related visual competencies are termed *visual thinking*. The director's objective demonstration of this knowledge creates the *videospace*, a term growing in usage, although still not as common as *creating the television image*. The term *videospace* allows the novice director to focus on the fact that television does not depict reality. It re-presents the object world. The television camera pickup tube and lens create a television image. It is the director who must create a television image worth our attention and this cannot be done by simply pointing the camera at the world. The director creates television space, or videospace, primarily through the use of dynamic composition and motion. To this we must add lighting, sound, props, and acting talent as other critical variables. To create interesting videospaces, the director must master traditional camera movements (See Figure 7–2).

Pan and Tilt

One major type of camera movement is accomplished by changes in the direction a camera can be pointed or aimed. With small cameras, a person can hand-hold the camera and move it left and right, up and down. Broadcast television studio cameras are usually too large and heavy to be held by hand; furthermore, hand-held camera motion is often jerky and disturbs audience attention. Studio cameras, therefore, are mounted on *camera mounting heads*, also known as *panning heads*. The camera and camera mount are then attached to tripods or tripod-related devices called *pedestal mounts*. The change in camera direction is achieved by the camera operator's use of a handle attached to the camera mounting head. This head allows for two types of movement: *panning*, which is horizontal movement as the camera is pointed from one side to the other, and *tilting*, as the camera is pointed up and down along a vertical axis.

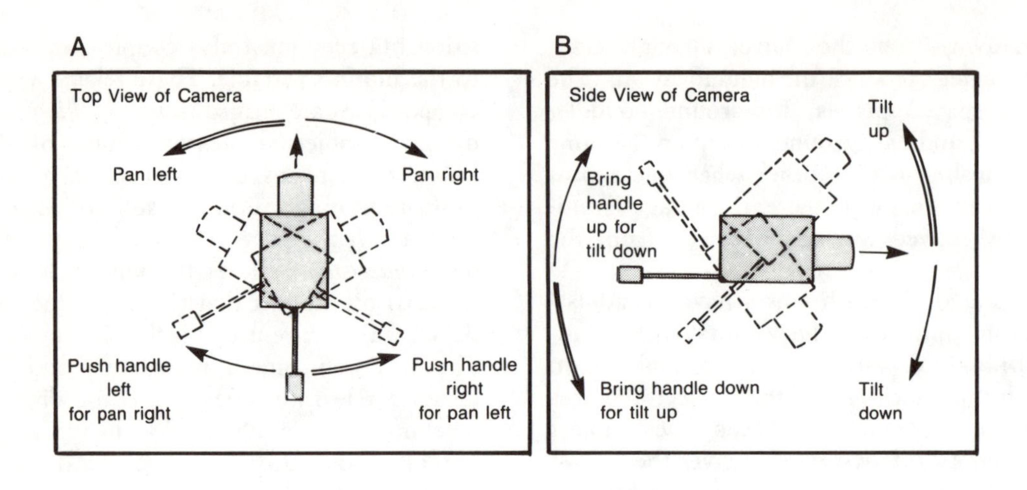

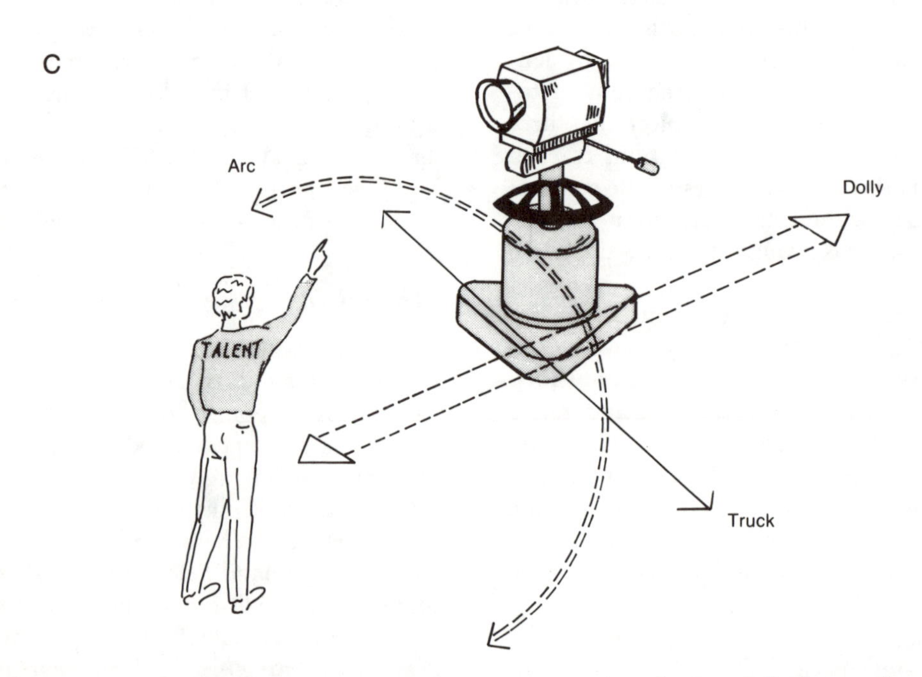

Figure 7–2 Basic camera movements. Pan: camera moves horizontally from a fixed axis (A). Tilt: camera moves vertically from a fixed axis (B). Truck, dolly, arc: camera axis shifts (C).

The television director communicates camera head movement change to the camera operator from the point of view of the camera operator. As the camera operator has the camera pointed at the subject, a new picture framing would be identified as being camera right or camera left (right or left of the operator). Thus a *pan right* or *pan left* command indicates that the director wants a horizontal move of the camera lens (via the handle attached to the camera head mount) towards the direction indicated. For vertical movements, the director commands the camera operator to *tilt up* or *tilt down*. Tilt movements can be combined with pan movements so that the net effect is a curved videospace. This type of curvilinear movement is rarely executed because of the difficulty of smoothly panning and tilting at the same time. To successfully achieve such camera movement, broadcasters often use the studio *cranes*. These are large camera mounting devices that allow the camera to be mounted on a *boom* or *tongue* so that the camera can easily sweep around the studio space. Cranes also have room for the camera operators. Cranes allow a camera to be moved left or right and up or down with superb fluidity. By combining both vertical and horizontal camera movement at the same time, the net result is a curved line. Camera cranes, however, are large and bulky. Many studios simply do not have the space to use a crane successfully.

Dolly-in and Dolly-out

A second major type of camera movement involves the movement of the camera along with its *tripod* (or *pedestal*). One of the most basic of such movements is simply towards or away from the subject. This is known as *dollying* the camera. To *dolly-in* means to move the camera towards a stationary subject or object. This type of movement is used when the director wants the audience to concentrate on a particular aspect of the subject without changing subject position in relation to the rest of the set elements. *Dolly-out* is the opposite of dolly-in. This type of movement is often used when the director wants the audience to first concentrate on a significant set element (object or action), and then to view this set element in a wider context. For example, a scene might open with a medium close-up of a young man breaking into a bedroom wall safe. The camera begins to slowly dolly-out and the audience recognizes (from previous scenes and other set cues) the bedroom as the young man's father's bedroom. Furthermore, the director could cue the father (actor) to enter the set as the dolly-out camera motion is being completed. This situation demonstrates how one camera movement can assist in the development or amplification of the narrative structure. Example: (1) young man stealing (director to camera operator, "Camera 1, dolly-out"); (2) young man stealing from father (director cues talent, "Enter stage left"); (3) father catches son stealing.

Walk-in and Walk-away

Related to dolly shots but not involving camera movement is a certain type of talent movement that is part of more complex camera/subject movements. *Walk-in* involves having a subject move closer to the camera. The scene may open with a waist shot (medium shot) of a subject. The subject then moves towards the camera while the camera remains stationary. The resultant shot is an MCU, also known as a shoulder shot. The *walk-away* is the opposite of the walk-in. For example, we may begin with an MCU and end with a wider shot because the subject moved away from the camera. We do not use the term *walk-out* because it has connotations which the director does not have in mind.

Truck and Arc

Truck and *arc* shots are lateral movements of the camera whereas dolly shots are towards or away from the subject. A director may command *truck right* and the camera operator(s) will move the camera towards the right along a line. The arc shot is a trucking movement of the camera along a curve rather than a straight line. It is combined with a panning motion of the camera so that the subject stays within the picture frame as the camera circles around part of the set. Arc shots can be highly effective but they require smooth integration of the arc movement with the pan movement.

Follow Shots

There are many types of *follow shots*. They have in common the fact that the subject and the camera are both in motion, not merely one or the other. One of the simpler follow shots is called *trailing*. For example, a subject may get up from a chair and start walking around the set in a prescribed manner. The director will give commands for the camera to follow the subject in a predetermined manner. The camera moves behind the subject, keeping a steady distance. This type of follow shot and all others require perfect timing and movement among performers and crew.

A more complicated follow shot involving trailing occurs in situations in which the director wants a new subject to be revealed to the audience as the next point of concentration. The camera initially trails subject A and then picks up a new subject B, thereby letting subject A fade to the background or entirely disappear. This switch in point of concentration can be done through a pan or a change in direction of the trailing camera movement. A number of combinations are available to the director but they must be judiciously used, especially as follow shots become increasingly complicated. This occurs, for example, in group situations when the director wants to make a transition from the individual to the group (or vice versa) without using the switcher to make a camera edit. Such follow shots usually require a number of practice *takes* prior to the actual recording of the scene. Despite their complexity, follow shots are highly dynamic when properly executed.

Camera Placement

Camera placement, or the positioning of the camera for a given shot within a sequence, is an important directorial decision. This decision affects not only how the individual shot will look, but also how it fits into the sequence that the director is constructing. Camera placement, thoughtfully or haphazardly handled, can make or break a program.

Several factors should be taken into consideration before making camera placement choices. First, what will occur within the shot and what are the important elements that must be conveyed? The director needs to make sure that all necessary action or detail can be covered from the camera position(s) selected. If the audience cannot see what they need to in order to understand the scene, camera placement will, at best, be ineffective, or worse, disorienting. Camera placement also allows a director to focus the audience's attention by excluding nonessential visual information. This minimizes audience distraction from the key components of the shot and allows the director to place more or less emphasis on any given item according to the degree to which it is isolated from its context.

A second question that the director should ask regarding camera placement is, "What is the mood or environment that I am trying to create?" Camera placement can aid the director in visually representing the "attitude" of the program. For example, if the situation involves two people conversing, the director may choose different camera placements according to the program's nature. Both individuals might be included in the same frame to give a feeling of intimacy to an interview, while for a debate they might be separated in a shot-reverse shot setup in order to add dynamism. Camera placement also can inadvertently influence how the shot is perceived if not given proper attention. The height at which the camera is placed is a notorious example: if placed too high, it can give the subtle impression of subject inferiority, or if placed too low, the subject may appear unduly powerful.

A final question for consideration is, "How does the shot contribute to the sequence—what shots precede and follow it?" Unless the director is after a special effect, the guiding principle of sequence construction should be the idea of *unity* and *clarity*. Correct camera placement can play a large role in achieving this by keeping the viewer spatially oriented and allowing the action to flow smoothly. When properly used, the following rules for camera placement will result in scenes that are both visually coherent and understandable to the viewer.

Jump Cuts

A *jump cut* is a jarring transition between two shots of the same subject. This undesirable effect occurs when there is a cut or dissolve between two similar shots on different cameras or between different framings of the subject taken from the same position. (The latter instance may happen if the scene is being *covered* for later editing.) For the viewer, it appears that the shot has remained the same except for an inexplicable jump in the image or that they are being jerked closer to or farther away from the image. Jump cuts can be prevented by making sure that there is an angle shift of at least 30 degrees between the two shots and/or by placing the camera(s) so that completely different shots are being covered. For example, cutting from a close-up on an oblique angle to a straight-on two-shot would avoid any jump cut problems; both the angle and the shot content have changed enough to make the transition visually acceptable.

Shot-Scale Transitions

Shot-scale refers to the director's choice of framing for a given shot (e.g., close-up, medium shot). Cutting between extremely different shot-scales can, like the jump cut, make for an extremely rough transition. Hopping from an extreme close-up to a long shot, for instance, does not allow viewers to spatially orient themselves in a smooth manner. Once again, the viewers are left feeling as if an invisible force is propelling them about. The rule of thumb here is to never cut to a shot that is more than three times larger or smaller than the previous shot. This guideline will help determine where to place cameras in order to have the necessary shot variety to avoid jump cuts, while not placing the cameras at such extremely different distances from the subject that shot-scale disparities become a constant headache for both the director and the camera operators.

Continuity

The notion of *continuity* is a key factor in achieving a clear and unified production. Continuity is the maintenance of consistent spatial and temporal relations with a sequence of shots. It aids the viewer in comprehending the visuals by constructing time and space in a logical manner. Without continuity, the viewer may easily become "lost in space" or temporarily disoriented through the presentation of disorderly and confusing visuals.

Content Continuity

One aspect of general continuity, *content continuity,* is concerned with the consistent positioning of subject/objects within the sequence. For example: there are two people on a talk show with a large chart between them. Both are occasionally indicating items on the chart. Camera one has a shot of person A on the left and the chart on the right of the frame. Camera two has a shot of the chart on the left and person B on the right. Cutting between those two shots would result in the chart appearing to jump from the right side of the frame to the left—a violation of the viewer's spatial logic. In this case, it would be better to pan from A to B, thus preserving spatial continuity, or to first cut to a shot including A, B, and the chart, which would also allow the viewer to remain oriented. Another aspect of content continuity comes into play when a program is being taped for later editing or if the taping is broken up over time (for example, the session is interrupted by a lunch break). In such cases, it becomes particularly important to make sure that costume, set, hairstyle, and other details remain the same. If someone moves a vase of flowers from one table to another or a performer forgets to put a coat back on, items may mysteriously vanish, move, and reappear between shots! The same applies to an action that must be repeated by a performer to permit it to be shot from a different angle; it must be repeated exactly.

Screen Direction Continuity

Screen direction continuity ensures that the direction of movement or subject placement within a sequence does not shift in contradictory or incomprehensible ways. As with the other principles of continuity, screen direction continuity is necessary for the production of a clear and unified sequence. Two major rules are used to meet this end. The first is that the direction of a single movement should be consistent across shots. If a performer is walking from screen left to screen right in shot 1, he or she should not appear in shot 2, walking from screen right to screen left. The same is true for any movement, whether it is a car, a ball being thrown, or a dog running. The second rule is that subject placement should not reverse between shots. If the talent is facing screen right looking at a guest, he or she should still be looking screen right if the director cuts to a two-shot. This does not mean that the subjects are immobile. It means that correct camera placement is vital to screen direction continuity. The so-called *180 degree line* is a device that has been developed for the determination of camera placement. To use this method, imagine a line drawn from the actors to the point they are moving toward (or if stationary, the point they are looking at) or, in the case of two subjects, a line which connects them. Now you have your 180 degree line, an imaginary line that divides a 360 degree space into two equal halves. One should simply keep the camera(s) on one side of the line or the other, but not cross over it. What will happen if the camera crosses the line? The most common version of this error occurs in a shot-reverse shot setup in which person A is shot from one side of the line and person B is shot from the other side. On screen, A and B will look as if they are looking in the same direction instead of at each other and the audience will want to know what they are both looking at.

Camera Angle

Related to camera placement is *camera angle*, an aspect of camera positioning that has the potential for fine-tuning the productions. Camera angle is the relationship between the camera's position and the subject, on both the horizontal and vertical planes. Skillful and creative use of camera angle can construct exciting shot sequences from seemingly mundane material. Camera angle, together with composition, camera movement, and continuity, can be the difference between an image that viewers perceive as worth their attention or one that is turned off.

On the vertical plane, the camera angle may be *high angle*, *low angle*, or *level*. A high-angle shot is one that looks down on a subject from above it; it tends to make the subject look smaller, weaker, and/or powerless. A low-angle shot is a shot in which the camera is pointed up at the subject; the subject seems to be much larger, stronger, and more powerful than normal. A level shot is one taken at a normal relation to the subject; it generally appears to be the same perspective an adult would see. The occasional use of high and low angles can add emotional depth (showing a child's viewpoint of an adult situation), comic relief (emphasizing extreme points of view in slapstick), or subtle character tones (consistently using a slight low angle on the "good guy," use of a high angle for the greedy banker's point of view of someone asking for a loan).

Camera angles on the horizontal plane do not have specific names due to the fact that the camera can be positioned anywhere from 0 to 180 degrees away from the direction in which the subject is facing. The one exception to this lack of individual nametags is the *Dutch angle* or *canted frame*. This is a special case, however, which is tagged not because of its spatial relation to the subject (it can occur at any position on the 180 degree line), but because of its unusual influence on the framing of the subject. The Dutch angle results when the entire camera is slightly cocked up or down; that is, it is no longer level. When this occurs the image likewise appears to be on a slant; floors, for instance, run diagonally in the frame rather than horizontally. This technique was made famous by the film noir movies of the 1930s through 1950s and was used to induce tension or suspense in the viewer. If overused, however, this technique can quickly become boring or completely disorienting.

In conjunction with composition and camera movement, camera angle can be used to influence the mood a director wants to project, to hide or reveal visual information, to add an aesthetic dimension to the shot, or to simply make the sequence more visually pleasing, as shown in the following examples.

Any type of extreme angle is going to be perceived by the viewer as more jarring; such angles, therefore, can be used to add melodramatic impact or to evoke feelings of anxiety, fear, etc. due to their theatricality. How many monsters, aliens, zombies, and ax murderers have there been who loomed over their victims (and the viewers) in an extreme low angle shot! Another creative use of camera angle can be seen in positioning the camera so that certain visual elements are obscured from the audience's view (perhaps a hat or long hair covers the character's face). This can increase the surprise when that information is later revealed and the audience realizes that the butler didn't do it! A completely different use of camera angle involves the enhancement of an image's aesthetic qualities by segmenting into various interesting views or by providing a view of the subject one might not ordinarily see. This can be used to good effect for dance or musical performances or views of nature, for instance. With narrative or informational programs it should be used sparingly so as not to unduly distract from the main focus of attention.

There are a number of generalizations about camera angles of special value to the director. Most of the time the camera should be positioned about 45 degrees away from the subject for a facial close-up. This position seems to be the most favorable for rendering subjects in a flattering manner and for giving depth and modeling to the image. Pairing an angle with its mirror image for a shot-reverse shot setup imparts a feeling of balance; if the close-up of A is taken from 45 degrees to A's left, the close-up of B should be taken from 45 degrees to B's right. The same is true for pairing over-the-shoulder shots, medium shots, etc. Of course, it is not always necessary to follow a shot with its mirror partner (a director may choose to move to a wide shot, for instance), but most of the time a director should avoid pairing a close-up of A, for example, with an over-the shoulder shot of B. It can throw a monkey wrench into the dramatic progression as well as fail to meet viewer expectations. The weakest and most static camera angle is to place it perpendicular to the 180 degree line so that the subjects are in profile. The closer the subjects are to facing the camera, the stronger the angle. Variety, however, is the best way to give the sequence energy.

Now that we have examined the primary elements that a director must consider when making camera placement decisions, we need to put those guidelines to work in practical situations, using *camera patterns* (i.e., combinations of camera positions). Knowing a variety of basic camera patterns will increase one's efficiency in dealing with frequently encountered situations and allowing one to adapt quickly to more unusual set-ups. It is important to remember that camera patterns should not only reflect the principles of continuity, camera placement, and angles as discussed above, but they also should serve to provide interesting compositional possibilities and be flexible enough to incorporate varied camera movements.

Camera Patterns

The two-person interview is a staple in television. It appears in news reports, talk shows, educational programs, sports features, just about everywhere. Needless to say, its standard camera pattern is both important to know and easily adapted. Generally, two people are either standing or seated facing each other. Instead of simply pointing the camera on the right at the subject on the right and the camera on the left at the subject on the left, the basic interview calls for *crossing camera angles*. In other words, the camera on the left shoots the person on the right and vice versa. This allows for better facial shots and also gives each camera the mobility necessary to get the two-shot or an over-the-shoulder shot of the individual it is focused on. As well as the director having several shot-scales to choose from, crossing camera angles multiplies the variety of available shots by placing each camera so that it can easily adopt three positions instead of just one. This same pattern also works well for a single individual speaking while simultaneously referring to charts or models, demonstrating a product, etc. It also could be used in a variety of other situations such as an individual signing for the deaf while speaking (one camera focusing on the face while the other focuses on the hands), a two-person dramatic dialogue, or a two-character puppet show.

A second situation that directors frequently encounter is the panel set-up. This entails the director covering several people lined up in a row facing the camera. Examples of this include the usual arrangements for anchorpeople on news programs, talk shows with multiple guests, panel discussion programs, and game shows. To facilitate adequate and varied coverage, three cameras are used, again with crossing camera angles. In this type of set-up the director not only needs to be able to get shots of each individual and of the entire group, but also of differing combinations such as two-shots or over-the-shoulder shots if the 180 degree rule will allow. Generally, one camera is placed centrally with an additional camera to either side at a 45 to 60 degree angle from the middle camera. Two of the cameras may then be free to adjust their positions while one of the cameras is responsible for a wide shot. In this way the action is always covered while opportunity exists for great shot variety. Another variety of this same set-up uses four cameras. In addition to the camera covering the wide shot, a camera continually covers the host or moderator.

The final pattern to be looked at is one used to cover a triangular configuration. This is probably most often seen in dramatic or comedic narratives and in musical numbers with multiple characters on the set simultaneously. The danger with this three character arrangement is that a continuity violation is much more likely to occur. To prevent this, the director must constantly be aware of the shifts in the 180 degree line according to where the line of action is for a given shot. As in the panel set-up, three cameras are used but they are deployed somewhat differently. In order to adequately cover all three characters without disorienting the viewer, it is useful to swing the two outside cameras over until they are at starting positions 75 or 80 degrees away from the central camera. In this fashion, good cross-angled individual shots and over-the-shoulder shots may be covered by the outside cameras while the central camera can easily cover the wide shot of all three or either two-shot. Characters may also "cheat" a little bit in the direction of the camera to improve the coverage. If the 180 degree line shifts with a change of emphasis in the action (e.g., two of the three characters begin to have an argument or sing a duet), the appropriate outside camera can continue arcing around to accommodate the change. (Figures 7–3 through 7–7 illustrate the use of the 180 degree and camera continuity adaptations.)

Visual Composition

As subjects move around a set, cameras have to be moved in various ways to maintain viable picture composition. For example, during a medium close-up of a seated subject, the subject rises up from a chair. If the director does not want the subject's face to appear "chopped-off," the director must give a *tilt* command, a *dolly-out* command, or a *zoom-out* command in order to properly frame the now standing subject. *Correcting composition* is a natural part of television production and should not be confused with correcting a mistake. Correcting composition is the traditional term used by directors

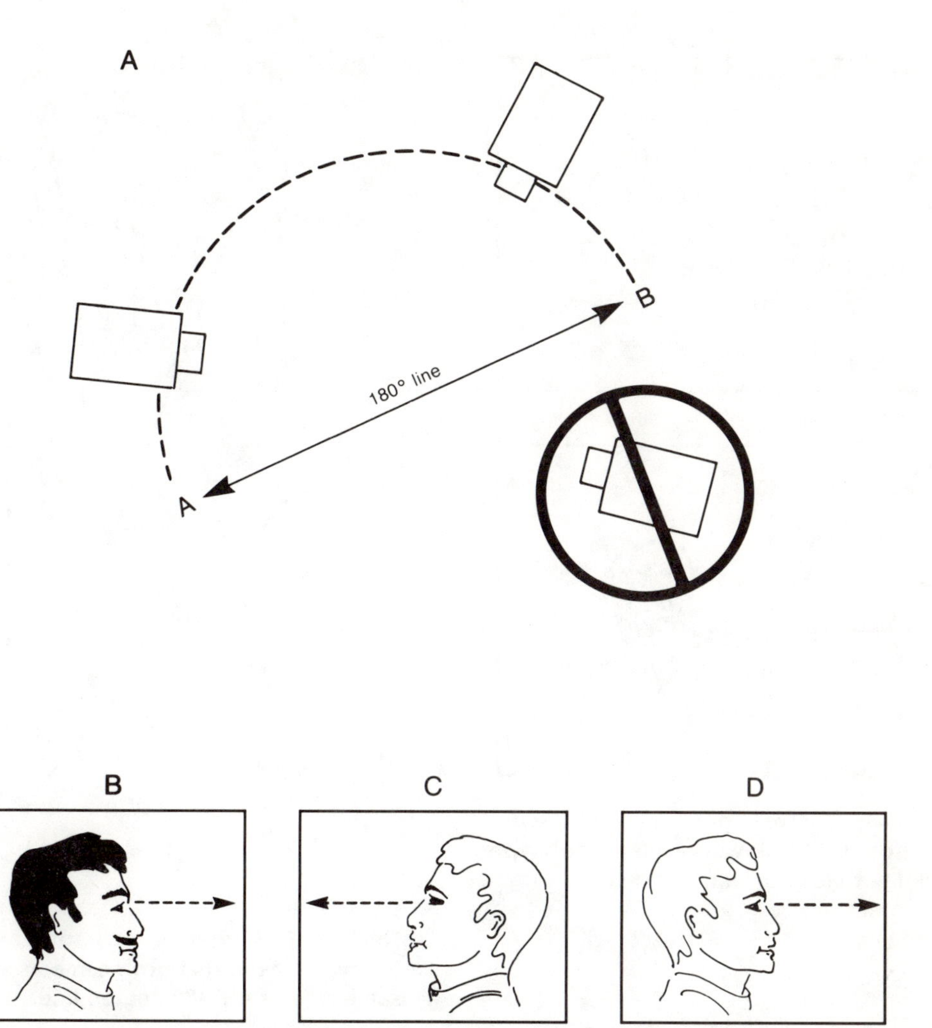

Figure 7–3 Camera placement and the 180 degree line. Shooting from both sides results in unclear spatial relations between images (A). Shot of Point B from one side of 180 degree line (B). Correct shot of Point A, taken from same side of 180 degree line (C). Incorrect shot of Point A, taken from opposite side of 180 degree line (D).

when they need to reframe the picture in order to maintain a balanced composition of the subject or subjects.

Many television directors first acquire a sense of visual thinking by learning the classic principles of visual composition as found in such traditional arts as drawing and painting. This is, of course, a good place to begin thinking about composition because certain similar arrangements of line, form, and color (among other elements) have seemed to please humans for thousands of years.

Harmony and balance within a pictorial composition, for example, have continual perceptual (and emotional) appeal for humans of almost all ages. Film and television production, however, requires modification of some classical principles of visual composition because they are motion media.

An aesthetically pleasing picture is composed of many artistic elements, including, but not limited to, *rhythm*, *mass*, *line*, *proportion*, *unity*, and *balance*. The last element is of utmost importance, especially for be-

Figure 7–4 Content continuity. Cutting directly from (A) to (B) will result in the graph appearing to leap from one side of the frame to the other.

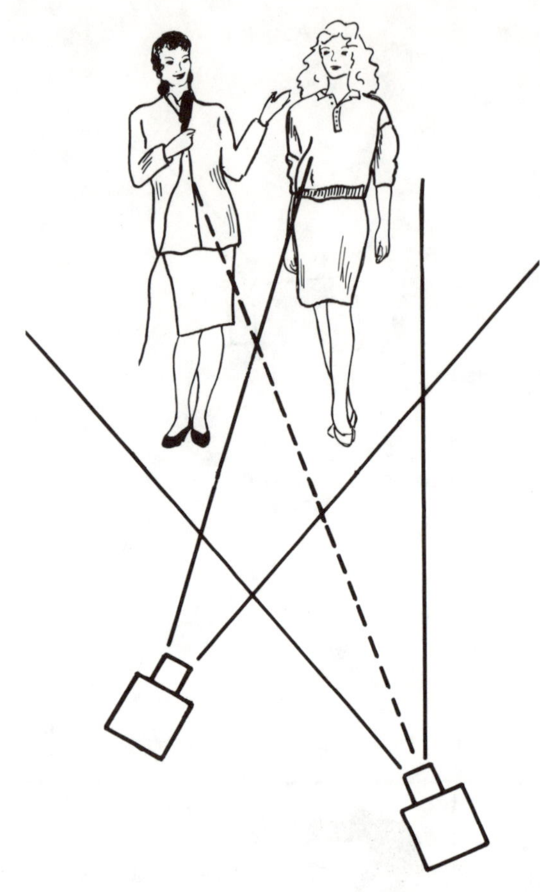

Figure 7–5 Camera pattern for a two person interview. Note that both cameras are on the same side of the 180 degree line.

Figure 7–6 Using three cameras to shoot a panel set-up. The two outside cameras can get individual or two-shots while the central camera covers the wide shot.

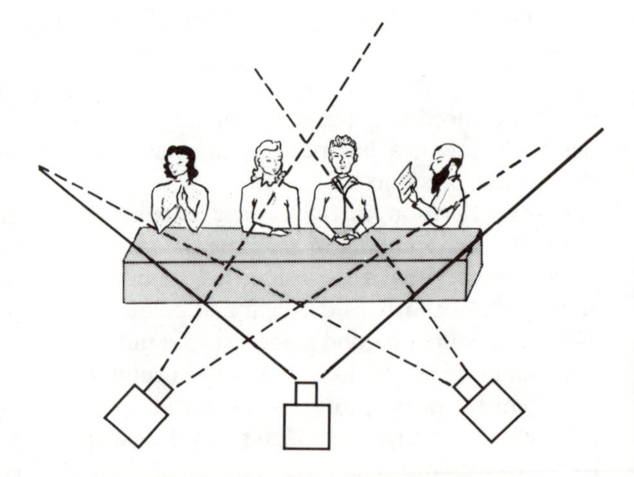

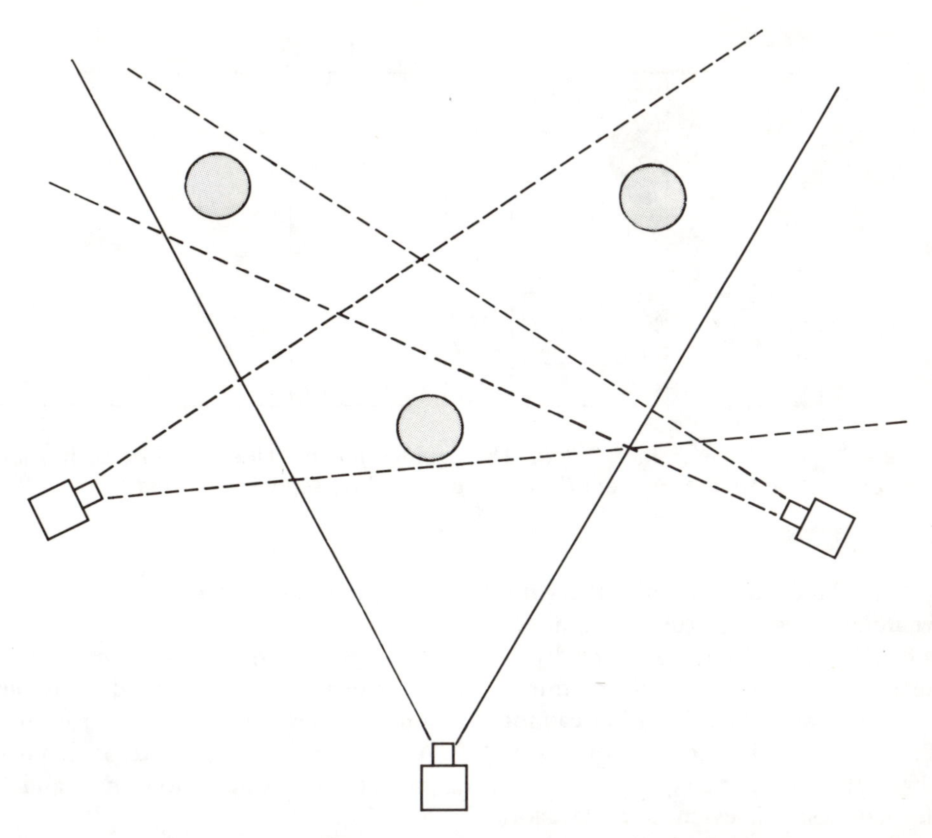

Figure 7–7 Triangular arrangement. Cameras must be placed so that they can easily get several different shots and accommodate shifts in the 180 degree line.

ginning directors who tend to strive for symmetrical balance within the frame. Symmetrical balance means the placement of the important picture elements in the center of the frame, or the placement of equal elements equidistant from the center. This gives the frame a nice even look, but robs it of tension; it appears static. A frame without tension is a frame that holds little attraction for the human eye. To avoid central, symmetrical pictorial composition, television directors use a visual principle called the *rule of thirds* (see Figure 7–8). This rule derives from the so-called golden mean principle attributed to the Greek mathematician Euclid. To visualize the rule, divide the television screen both vertically and horizontally into equal thirds. The intersections of the lines are said to represent the aesthetically perfect focus of interest. Television directors place the major pictorial elements at the point where the lines intersect, especially the left-upper and right-upper intersections.

Of course, a television director can use the rule of thirds successfully and still produce a rather dull or formal pictorial composition. Don't forget that the rule of thirds assists us in developing excellent still picture composition. But in a motion medium, excellent composition is necessary but not sufficient for the creation of dynamic images. For example, a director might have each sequence perfectly composed according to the rule of thirds. This director then places the camera directly facing (or *straight on*) the talent. The resulting picture will appear very flat on the TV screen even though it has balance, weight, and harmony. Each aesthetic

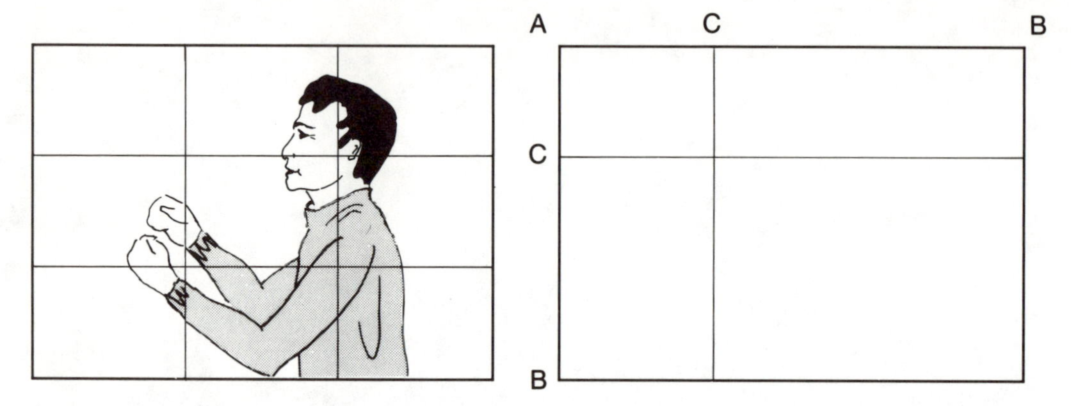

Figure 7–8 The "rule of thirds" *(left)*. The "golden mean." Line segment AC has the same proportion to line segment CB as CB does to line segment AB *(right)*.

principle must be combined with others for truly dynamic television pictures. The director here failed to combine the need for depth with the rule of thirds. The rule of thirds helps us towards achieving depth but cannot achieve it alone. (See Figure 7–9 for examples of balance and symmetry.)

The formal, aesthetic essence of television production is to achieve the feeling of three dimensions in a two-dimensional space. To do this, directors experiment with camera placement, camera angles, and subject/camera movement. Directors rarely use straight line arrangements; they prefer the triangular, the semicircular, the circular, and other irregular patterns. Of course, lighting is critical for the development of depth and most television directors have a good grasp of lighting principles even though they hire lighting designers who specialize in that aspect of production.

Directors use many other compositional rules to achieve dynamic, pleasing television pictures. But successful directors also know that the purpose of all these rules is to go beyond exciting, pleasing composition: to place the pictorial elements in such a way as to achieve an appropriate empathic response. The audience needs to be drawn into the production, to have some emotional connection with the piece.

Transitional Devices

In addition to composition and camera movement, the successful director needs to join together effectively sequences of images. A number of accepted transitional devices move us from camera to camera and/or scene to scene.

The Cut

The quickest, simplest, and most frequently used method of transition from one camera to another is the direct *cut*. This is accomplished via a black box called the camera switcher. Cutting from one camera to another is usually motivated by the desire to show the action to the viewer from different angles or points of view. A cut can also be used to join or relate scenes. Most transitions are straight cuts. Directors usually follow the *3-2-1 rule*: never cut from a wide shot to a close-up without a medium shot inserted between the two.

Traditional uses of cutting are directed towards the manipulation of space and time. By cutting from a medium shot to a close-up, for example, the director creates a dynamic visual impact upon the viewer. The proper use of cuts for movement through space depends on an excellent sense of visual composition. The director must relate the

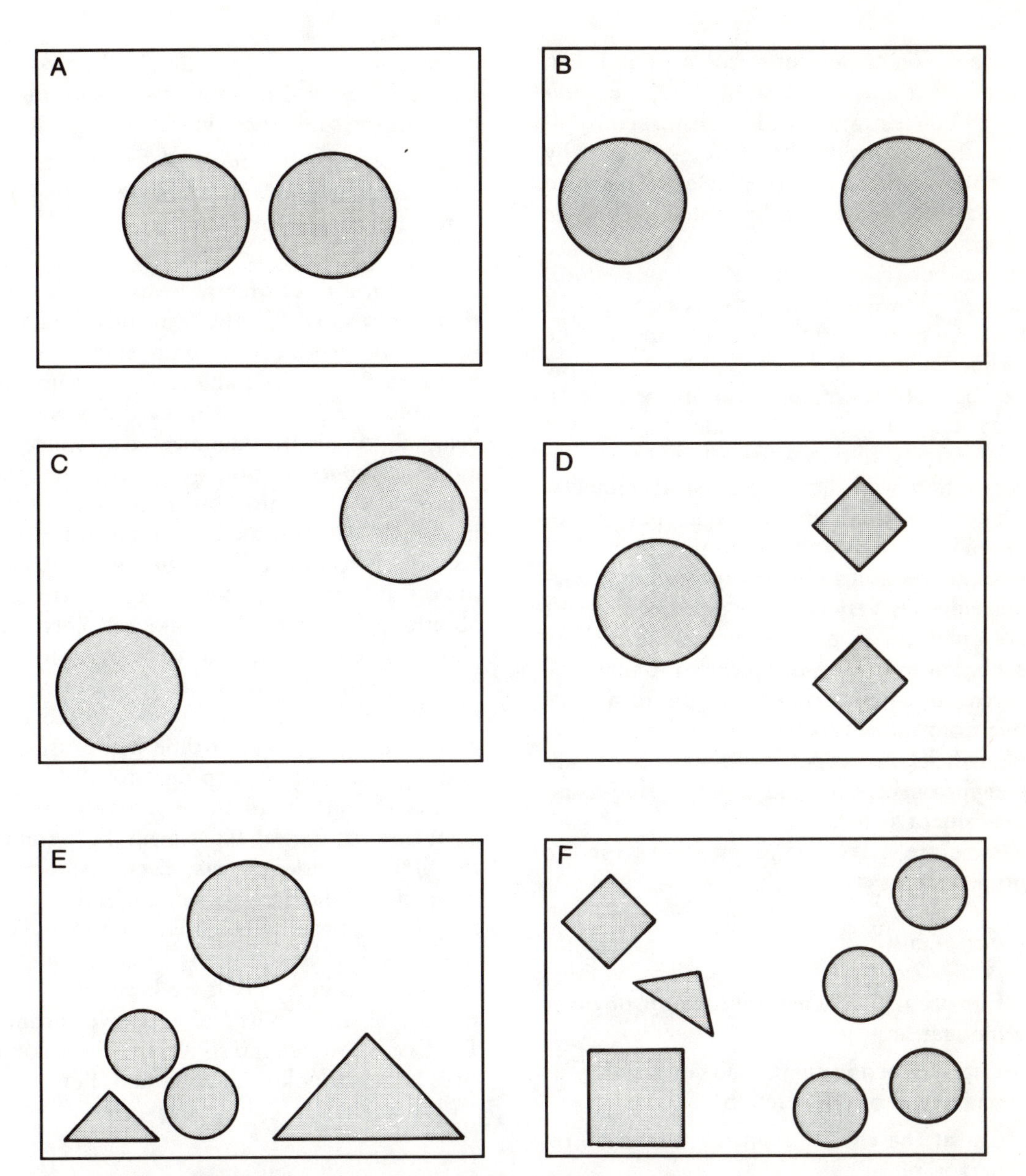

Figure 7–9 Effects and examples of balancing and symmetry within the camera frame. A symmetrically centered composition appears static (A). An asymmetrical balance of two elements leads to a more dynamic composition (B). Magnetism of frame edges. Note that subjects are pulled apart, leaving excessive "dead" space in the center (C). An asymmetrical balance of three elements (D). An example of balancing multiple elements within a composition (E). Another example of combining various elements to form a balanced composition (F).

composition of one cut to the next and then to the next and so on. Otherwise, the audience's concentration will be disturbed by illogical spatial relations and/or deflected by inferior aesthetic values that arise between frames (not within a single frame). The rate at which the director cuts from one camera to another (*cutting rate* or *cutting rhythm*) is used as a means to express concepts of time. An increase in the cutting rate gives the feeling of increased tempo. This technique can be used to create a variety of moods, such as anxiety, fear, excitement, and joy.

Another traditional use of the cut is to correlate it with dance or musical numbers. The syncopation and/or asyncopation between the visuals and the sound can be used to create tension and release within the viewing audience. Very complex pieces of this sort are often cut in postproduction during the editing process. The complexity prohibits the successful completion of such pieces by cutting from camera to camera via the switcher. Nonetheless, an experienced director can accomplish successful and rapid cutting using the camera switcher.

Here are a few major principles for the proper use of cuts:

When to cut:

Cut on action, when there is an obvious motivation.

Cut on reaction: make the cut a split second ahead of the reaction.

Cut at the end of a phrase, sentence, or paragraph.

Cut for dramatic effect with different angles for variety and movement.

Cut at the end of a musical phrase.

When not to cut:

Do not cut in the middle of a sentence.

Do not cut to the same shot on a different camera.

Do not cut from a pan to a stationary shot.

Do not cut in the middle of a pan shot unless the cut is to another pan shot moving in the same direction.

Do not cut just for the sake of cutting; always plan cuts and be aware of the motivation for the cut.

Television directors are aware of the fact that a cut is an abrupt transition and, by definition, produces a momentary disturbance in the viewer's ability to concentrate. The rules for proper use of cutting were developed, in part, to minimize disturbance of audience concentration—to mask the abruptness of the transition. Experimentation in the 1970s confirmed what successful directors always believed: that a poor choice of cuts disturbs the audience's concentration abilities. Certain types of illogical cuts can cause audiences to lose concentration for as much as ten seconds before they can re-attend to the image. Too many unsuccessful cuts in a short period of time will cause the audience to turn off a program. There are always exceptions to these generalizations, as in cases of special experimental programs for film and video artists. Even traditional dramatic productions sometimes violate cutting rules in order to heighten emotional tension. Nonetheless, it is important to remember to keep a balance between excessive use of static shots and excessive cutting. The first is too flat and show, and the second gives a jumpy look to the program. Both turn off the audience.

The Fade

The term *fade* indicates that a major change is being made in the continuity of the show. A *fade-in* is a gradual transition from black (no picture) to a picture of full intensity. A fade-in often is used to open a television program. This is similar to the curtain rising in a theatre production. A fade-in frequently is used after commercial breaks. Fade-ins are used also to signal a major time shift or to indicate a change in the story line. For dramatic effect, slow fade-ins are often used.

The *fade-out* is the reverse of the fade-in; that is, the picture fades from full intensity to black. Fade-outs are used to enter commercial breaks, and to signal the end of a time period, the end of a story line, and the end of the program. A fade-out (except at the very end of a program) can lead to audience anticipation of a new story line or a new period of time within the narrative.

The rate of the fade conveys information. A very rapid fade softens what would otherwise be a cut. A less rapid fade conveys a sense of drama and suspense. A long, slow fade often conveys a sense of peace and calm. At the beginning of the program, the rate of a fade-in can set the initial tone for the first sequence. In designating the passage of time, the length of the fade is proportional to the amount of time that has passed.

The Dissolve or Mix

The *dissolve* is a blended transition; it is a cross or double fade. One picture source is faded out to black while another source is faded in to a picture. Just prior to the beginning of the dissolve, picture 1 (i.e., camera 1) is at 100% intensity; this is what the audience sees. Picture 2 (i.e., camera 2) is at 0% intensity (black). As the dissolve begins, picture 1 starts fading out towards 0% while picture 2 starts fading in towards 100% intensity. At the midpoint of the dissolve, each picture is at about 50% intensity. For a split second we have a *superimposition*: two pictures on the screen at the same time.

Dissolves often are used to indicate a brief passage of time or perspective. This is similar to the effect of a fade except that the dissolve usually indicates a short passage of time, or a minor change in locale or story line. Shot selection, choice of camera angles, and changes in movement must be given proper attention for a dissolve to succeed. Rhythm is also important with the use of a dissolve. The rate of a dissolve should fit the desired tone of the program and it must aesthetically flow from shot 1 to shot 2. If a major dramatic shift or shock effect is desired, the director will use a cut instead of a dissolve.

A *matched dissolve* is a transitional device that involves two cameras with shots of closely related subjects or objects. It differs from the ordinary dissolve only in the relationship of the two shots.

The Wipe

A *wipe* is an effect made possible by black boxes called special effects generators. The camera switcher and the special effects generator (SEG) are contained in one box. Wipes can look quite different, depending on the particular machine available to the director. An example of a wipe is the *split screen* effect. This usually takes the form of a vertical line that divides the screen into two equal halves. The dividing line itself can be a hard line, a diffuse line, or a line with a color border. The use of this wipe is quite common on talk shows. One person may be in the audience and the other on stage. Rather than continually switching back and forth between the two people as they talk to each other, the director will use the split screen wipe effect so that the audience can view both people at the same time.

Many other wipes can be called pattern dissolves. A new pattern is introduced into the current image on the screen. The new pattern will grow and finally supersede the original picture; in fact, the pattern itself often disappears because it soon takes up the entire screen and thus looks like a normal camera shot. To take another example in our talk show: shot 1 is the full screen of our guest talking to the host. The host then faces the studio audience and solicits questions. Camera 2 focuses on an audience member indicated by the host. As the studio audience member begins to speak, the director decides to use a *corner wipe* by engaging camera 2 as well as camera 1. The viewing audience sees the studio audience member talking in the bottom right (or any other corner) portion of the screen. This effect can start out with just a 2 inch square in the bottom right corner, and as the studio audience member keeps talking, the director can continue to wipe-in the camera 2 shot so that it takes up an increasing area of the frame—even until shot 1 (from camera 1) entirely disappears. The director simply could have cut to the studio audience member rather than attempting this special wipe effect, but used it for aesthetic and psychological purposes. These kinds of wipes are not frequently used, but directors use them to give their shows a special look and to increase audience interest through unusual imagery.

Defocus to Refocus

This is a simple transitional device accomplished through manipulation of the camera lens. It is usually achieved by defocusing the lens of one camera while another camera lens is already defocused and ready to be brought in by the director via a dissolve effect. Since both cameras are out of focus during the dissolve, the viewer is never aware of the dissolve technique, but only aware of the defocus. As soon as the dissolve is executed from shot 1 to shot 2, shot 2 is brought into focus by focusing the lens on camera 2. This technique often is used to convey a sense of time passage as in a flashback, a dream sequence, or loss of consciousness. The defocus technique frequently is used in pairs to indicate the beginning of a flashback and the return to present time.

Contemporary special effects generators have so much special image capability that directors are using other techniques in place of the defocus to refocus technique. This adds variety to the program. One such technique is *digitalization*: the screen is broken into hundreds of little squares so that the picture looks like it is a low-resolution computer graphic.

The Super

This term is derived from *superimposition* and is comparable to the effect of double exposure in photography. It is a means of placing one image on top of another and is rarely used with two full-frame images. Instead, it is commonly used to roll credits on top of a camera shot. One camera will be focused on the studio set and one camera on a special machine that prints letters in white ink on a black card or roll. Both cameras are on and the director selects superimposition on the video switcher. Special effects generators accomplish a similar look with more variety and precision with *key in* or *matte* effects.

THE CONTROL ROOM

The control room is often called the heart, the brain, or the nerve center of broadcast television. Larger stations may have two or three control rooms connected to two or three studios. Control rooms will look somewhat different depending on the station, the period in which they were built, the designer's plan, and later modifications. For example, during news programs the news anchor may inform the audience that it is going to hear Senator Y's opinion on issue X. Just as the anchor finishes speaking, the director, via intercom, informs the videotape player operator to "roll segment 1." The director then switches to tape playback via the video switcher. Where are the tape machine and tape operator located? Some control rooms have the VTR playback machines located within them. Other control rooms can be a fair distance away from the location of the tape playback equipment. In some control rooms, the video director can see the television camera controls; in other control rooms, the camera controls are placed in front of the engineers who are in an adjoining area. Despite these design differences, control rooms always include certain types of equipment (see Chapter 4). How does a director use these tools?

From the director's point of view, the most important piece of equipment in the control room is the video switcher, which controls all picture aspects of the program. The director is positioned directly in front of the switcher or slightly to one side. If the director has an assistant, called the *technical director* (TD), that person will actually operate the switcher as commanded by the director. The switcher is built into something like a tabletop. It is often tilted upwards, towards the director's eyes. This makes it easier for the director to look both at the switcher and the television monitors positioned above the switcher. There are many monitors in a control room. There is a TV monitor for every camera so that the director can see what the cameras are shooting. There are monitors for tape playback, for character generators (machines that print out words and numbers for graphic needs), for the film/slide chain (so the director can see what those images look like), for the *line* (what actually goes out to the home audience), and for preview (more about that below). In short, there must be a TV monitor for every electronic image that the director may use. In addition, there are an intercom and a speaker (or headphones) for communication with all other production personnel no matter where they are located. The control room may also contain various types of audio equipment (turntables, audio tape recorders, etc.) to support the program's audio needs. A television director usually has audio staff who operate the equipment at the director's commands. In addition, control rooms also may contain graphics equipment, videotape playback equipment, film/slide chains, etc. But not all control rooms will physically house this much equipment; adjoining areas are used.

The Video Switcher

The video switcher controls all aspects of the image. It can be a simple device with a small number of buttons and levers, or it can be an expensive, complex device with more than one hundred buttons. Each switcher will be different depending on manufacturer and cost, but they all provide five main functions: (1) they allow the director to choose different camera shots by switching among cameras; (2) they provide the director with special effects such as fading, dissolving, superimposing, and keying; (3) they allow the director to switch to noncamera images originating in other sources such as VTRs and film/slide chains; (4) they allow the director to preview image selections in advance; and (5) they deliver the director's final image choices to the line for broadcasting. (See Chapter 5, Figure 5–12, for an illustration of a state-of-the-art video switcher.)

To control all these functions in a logical manner, video switchers are designed with the switching buttons aligned in parallel rows called *banks* or *buses*. The simplest switcher usually has a minimum of three banks or buses. One row of buttons is the *line* bus. There is a button for every possible image source (e.g., cameras 1, 2, VTRs 1, 2, film/slide chain). When the director depresses one of the line buttons, that image is selected for broadcast. Located above and connected to the switcher is the line monitor so that the director can observe what the audience is viewing. Another bank of buttons is identical to the line bank but it is called the *preview* or *preset* bus. This bank is connected to a preview monitor located next to the line monitor. By *punching up*, or depressing a button on the preview bank, the director can look at the image that is going to follow the current image being viewed by the audience. Since the line and preview monitors are next to one another, the director can compare the current image and the next one. Adjustments can be made prior to actually sending the next image to the audience. This is especially important with special effects since they can be difficult to control and the director needs to preview them prior to actual transmission. If a director is satisfied with the preview image, he or she then punches up the same button (e.g., camera 2) on the line bank at the appropriate moment, sending it to the audience. Most switchers today allow the line and preview/preset bank to alternate positions; the line bank can become the preview bank and the preview bank becomes the line bank. A special separate button called the *cut button* performs this task. So if the director likes the preview shot, he or she can punch up the cut button and the preview bank becomes the line bank. Sounds a bit confusing? It takes experience to effectively use a switcher to its full potential. The cut button provides the director with more efficiency and flexibility.

If a video switcher is to provide the director with special effects, it needs to have a minimum of three more banks. Two of these banks, called the *A* and *B banks,* have almost all the same buttons as the preview and line banks. This allows the director to perform various special effects, some of which we have already discussed. A director who wants to superimpose camera 1 onto camera 2 punches up camera 1 on one bank and camera 2 on the other. To actually select the superimposition, a third bank, the *effects bank,* is needed. This bank contains the special effects choices provided by that particular switcher. The director previews the special effects choice and then punches up the effects button on the line bank when it is time to send that particular special effects image to the audience.

Why is there a need for three banks to perform a special effect? It helps to think of the effects bank as the command or decision bank. Its buttons represent the final image desired by the director. Is it a mix, a wipe, or a superimposition? But for the effects bank to decide something (that is, the director's punching up an effects button decides), it needs something to decide about. To choose *mix* on the effects bank means that the effects bank has to mix two or more images. So the A and B banks are the primary electronic source banks for the effects bank. When the director punches up buttons on the A and B banks, that information is sent to the effects bank. When the director punches up an effects button, the effects bank now has imagery to effect. Without that information, the effects bank cannot operate because it would not have any electronic sources with which to operate.

State-of-the-art broadcast switchers have many more buttons and banks than those described so far. For example, they have *fader* bars that allow the director to fade in and out from one image to another. In addition to the A and B banks, there may be a bank of buttons for wipe effects. This allows the director to choose various electronic patterns prior to punching up the wipe button in the special effects bank. There are also *key* buttons. This allows for another type of special effect (described below).

No matter how complex a switcher becomes, its complexity is for two reasons: (1) to allow the director greater control and efficiency in the production process and (2) to give the director access to a greater variety of imagery. Not all special effects imagery may be built into the video switcher. It would make a switcher too large and difficult to control. Digital effects and *painting* effects may be housed in separate machines, which are then connected to the switcher. An operator works these external machines and the director selects the images they send to the switcher at the appropriate moment.

Keying

One special effect extensively used by directors is the *key* effect. It has largely replaced the superimposition technique, especially for titles. Unlike the superimposition, in which one image is simply placed over another, the key effect enables the switcher to cut an electronic pattern into one picture and introduce a second video source into that area. The keyed effect produces a stronger, crisper image than that obtained by a superimposition because a super mixes two video sources, thereby compromising the luminence strength (brightness) of each image. For example, white letters supered over a person's chest allow us to see through the letters to the person. White letters keyed over the person's chest will be opaque and solid. This effect is called black and white *internal keying. External keying* produces the same electronic effect, but it uses an additional third video source. *Chroma keying* produces in color the same net effect as black and white keying although, from an electronic engineer's point of view, it is a bit more complex an effect. The keyed portion is referenced against hue rather than luminence (brightness scale).

Keying has given directors great flexibility in terms of producing images that look superb to the audience but do not actually represent a physical reality. For example, weather anchors stand in front of huge weather maps that move and change colors and give us all sorts of information. That weather map is an image on a TV screen being shot by a camera using a close-up. The weather anchor is then shot (usually an MCU or MS; it changes during the report for purposes of variety) with another camera. The director then keys the weather anchor into the TV-map shot. There is no such thing as the 8 foot tall moving weather map! But it looks to us as if there is. You can imagine the visual world opened up by the key effect. Without the key effect many images we take for granted, especially on TV news programs, would not be possible.

There are some limitations with the key effect, especially with the use of the chroma key. The director has to carefully match the colors of the sources being used. To chroma key a newscaster into a blue background would require the news anchor not to wear any blue items. The chroma key references hue and reads the dominant color of the portion of the image into which we wish to make our insertion. If the anchor wears a blue suit, the switcher will read that as the background color and the anchor's body will disappear except for the head and neck. There are other problems with chroma key involving image stability and proper lighting. The effect can look unpleasing if not properly controlled.

Potential versus Utilization

The potential of video switchers and special effects machines is much greater than their actual use. The development of computer-generated imagery and digital television has opened up entire new dimensions. Many of the new electronic image-processing tools have more than one hundred buttons for controlling the video image. Computer videographic systems and digital optic systems have thousands and thousands of image possibilities. The television director simply cannot control so much potential image creation; therefore, additional staff operate these tools and *ready* pictures for the director. These processing tools are connected to the video switcher so that the director can punch up those images as needed.

Audiences have not yet seen the image potential of these video processing tools because we do not have a proper context in which to introduce the new imagery. Some of the new graphics are seen in news shows, sports telecasts, advertisements, educational programs, music videos, and experimental programs. But prime time dramatic television does not readily lend itself to image experimentation.

THE DIRECTING PROCESS

Before the tape ever rolls for a studio production, the director and the production team must be thoroughly prepared. Adequate preparation assures the identification of potential problems and their solutions, as well as provides an opportunity for new creative insights. Preparation also helps to reduce costs, meet deadlines, and give the crew confidence, all vital to efficiently producing a quality program. It is the director's responsibility to map out the plan to be followed in meeting these goals. Usually this process can be divided into three distinct phases: script analysis and prepration, production rehearsal, and a check-out procedure that immediately precedes production.

The Script

The script is the blueprint that provides the director with the information needed to construct the specific visual and aural components of the program in an organized manner.

Using the script as a guidebook, the director decides how the production will proceed from beginning to end. This procedure is a somewhat complex one; the director not only must consider shot selection and camera placement, but also lighting, microphone placement, talent movement, props, sets, costumes, music, sound, and additional visual sources such as slides, charts or films. Depending upon the type of program, any or all of these elements must be planned in order to ensure proper execution.

Script formats may vary from station to station and according to the program's content. Generally, however, all scripts contain either specific dialogue and action or a description of what is expected to occur. Either way, the script provides the director with an easily discernable indication of both the audio and video content of the program. After reviewing the script and becoming familiar with it, the director's work really begins—each detail must be visualized and heard in the mind of the director before it can be concretized in the studio.

Marking the Script

The script serves as the director's map for the production, but the director must add road signs to show the exact route to be used. This procedure is called *marking the script*, and each crew member receives a copy with the director's notes. A marked script includes short notes and standardized symbols that indicate cues for sound, music, or talent; for timing; for the shot or shot-transition to be used in a given spot; for camera movement; and for special effects. These notes help guarantee that nothing is forgotten or overlooked during the excitement and tension of the actual production. On the other hand, a director should not *overmark* a script because clutter can confuse rather than clarify; notations should be used with discretion. One technique favored by many directors as an aid in interpreting their markings is to color-code them. For example, a red pen is used for camera 1, a blue pen for camera 2, and a green pen for camera 3. Audio cues usually are underscored or circled and written in a different color of ink to distinguish them from visual elements.

The following are some of the most commonly used abbreviations and their meanings.

T-1 (Take 1) Switch to camera 1

DISS-1 (Dissolve 1) Dissolve to camera 1

DI (Dolly in) Move camera in closer to subject

DB (Dolly back) Move camera back to wider shot

Tight CU (Tight close-up) Very tightly framed shot of subject

Loosen up (Wider shot) Usually given when camera is off air; a change of lens is in order

CU-LS (Combination shot) Shot of two subjects, one seen close up, one seen in distance

OS (Over the shoulder) Shot taken over the shoulder of one subject that includes another subject or an object that the subject is looking at

SUP (Superimposition) One shot is layered on top of another shot at this point in program

F-I (Fade-in) Scene gradually increases in intensity from black to full image; can also apply to increasing audio level

F-O (Fade-out) Scene gradually decreases in intensity from image to black; can also apply to decreasing audio level

Pan-L (Pan left), Pan-R (Pan right) Camera moves left or right on horizontal axis

Tr R (Track right), Tr L (Track left) Camera and dolly moved horizontally to scene

Tilt Up, Tilt Down Camera moves up or down on vertical axis

FL (Flip card) Cue floor manager to flip card

SL (Slide) A slide is to be used at a given time in the program

CUE or Q (Signal) Give signal to talent or crew member

THEME (Music) Indicates that music should be brought in

This is not intended to be a comprehensive listing, but rather to give an indication of how a director might abbreviate. Each director develops an individual system. The important thing is to be consistent so that the crew understands. These abbreviations often appear in combination, such as C-2DB, FI Theme, which means "camera 2 dolly back, fade in music."

Rundown or Routine Sheet Script

When a semiscript is used (i.e., a script that is fairly detailed but not word for word as it will occur), which may include production instructions from the writer, the director still marks it up. The familiarity of one's own markings makes sure that nothing is overlooked. Occasionally, a director may be called upon to direct from a *routine* or *rundown* sheet. This is a detailed script outline used with shows that have well established formats and patterns. The rundown sheet gives a descriptive rather than dialogue rundown of what will occur in a segment allotted for that activity. While this approach leads to some visual spontaneity, the director's selections are not random. Both director and crew know the routine and stay on their toes.

Timing, as in any production, is crucial. The following example shows a rundown sheet for a sixty-minute variety show. The left hand column is clock time or real time, the right hand column is the exact timing for each segment and the place where it will occur.

11:07:30	Allen welcomes guest #1	(10:00) Center Area
11:17:30	Allen CUEs	(0:30)
11:18:00 F-I film	Computer COM (commercial)	(1:00)
11:19:00	Jazz quartet plays	(3:00) Band Area
11:36:30	Allen intros Stacy	(1:00) Center Area
11:37:30	Stacy welcomes guest #2	(6:00) Corner Table
11:43:30	Allen gives newsbriefs	(5:30) Center Area
11:49:00	Allen intros	(0:30) Center Area
11:49:30 F-I Film	Aspirin COM	(1:00)
11:50:30	Stacy introduces guests #3 and #4	(6:00) Corner Table
11:56:30 SL	Allen does Cola COM	(1:00) Booth
11:57:30	Stacy wraps up show	(0:30) Center Area
11:59:00	Jazz quartet plays	(1:00) Band Area
11:59:00	SUP credits	(0:30)
11:59:30	F-O PIC (picture) and THEME	

Production Rehearsal

All directors complain about inadequate rehearsal time. If we were to imagine an ideal model of adequate rehearsal time, it would be something like the model below.

Similar to the initial rehearsals of a stage play, but with less time in television, the director goes over the essence of the script: aesthetic goals of the play, the theme, plot line, characters, overall sense of timing and rhythm, style, and, with each performer, an analysis of the character each plays, including physical and vocal characteristics as well as character motivation.

The professional director comes to rehearsals with a well marked script, knowing in advance exactly what each shot and each movement will be. Following the initial rehearsals, the director marks the floor(s) and wall(s), usually with tape, to establish prop and character locations. The director then moves the performers through the shots as they have been visualized and marked, making changes that seem necessary. For each shot, the performers are told where the cameras will be and what kind of shot will be taken. Directors use different approaches. The great television and film director Sidney Lumet always came to his TV show rehearsals thoroughly prepared, and frequently would hold a water glass up to one eye, pretending to be the camera as he put performers through each movement and shot he had marked in the script. Because crew time is expensive, a good director tries to get everything perfect in these *dry runs*.

Finally, the director moves cameras into position to confirm that shots will work as planned. The next step is a full camera rehearsal, going through the show without stopping. The main purpose of this rehearsal is to thoroughly confirm the effectiveness of the camera shots and continuity, and to test the overall timing of the show. Finally, the director calls for a full dress rehearsal. At this time any further modifications are made. The above schedule is an ideal. Few directors get so much rehearsal time.

The Check-Out Procedure

No matter how much or how little time is available for rehearsal, the director is always responsible for a complete preproduction check-out procedure. The purpose of this check-out is to make sure that all elements are present and properly functioning. For example, are the cameras placed correctly and the camera operators standing by? Are all mikes functioning? Are the light levels set correctly? Basically, the director must be sure that each individual crew member is in place and that equipment is working at optimal levels. In addition to the camera operators and lighting technicians, the director must check on the VTR operator, the audio and video engineers, the floor manager, film and slide operator, special graphics operators, talent, and any additional staff.

The Production Moment

How is an actual television production carried out? First, in a major broadcast television production, the director requires at least one, if not two, trusty assistants. The first assistant is the floor manager. This crew member is not in the control booth with the director. As with the camera operators, the director communicates with the floor manager via intercom. The director gives the floor manager advance warning of all action so that the floor manager can make sure that the talent is prepared and correctly positioned. The floor manager also communicates to talent any problems that the director notices, such as not speaking loud enough, walking into a shadow area of the set, or facing the wrong camera for an introduction. It is the floor manager's responsibility to check the set, to relay timing cues to the talent, and to indicate which camera is taking the shot. The floor manager can also aid the director in making sure the program times out correctly by signaling the talent to *wrap it up* or stretch something out. Standardized hand signals are used to communicate this information because the floor manager is on the set and cannot communicate verbally (see Figure 7–10).

The second trusty assistant is the technical director (TD). The TD sits adjacent to the director and in front of the switcher. The TD's job is to punch up all switching commands, from simple cuts between cameras to fades to special effects. Every switcher operation demands a split second of concentration on that operation. If the director shifts eyes to the switcher for that split second, it means not being able to observe any other function. There may be six switches in ten seconds. Without a TD, the director is unable to observe all operations on a continuous basis, with a loss of control over the production. With a TD available, the director can continue to monitor all production functions while calling out cue commands to the TD. The director also must issue all script commands as well as watch the camera monitors, preview monitor, and line monitor.

The director is responsible for maintain-

Figure 7–10 Standardized floor manager hand signals for communicating during production. 1. Stand by . . . go ahead. 2. Speed up, faster pace, quicker tempo (speed according to increase required). 3. Slow down, slower pace, stretch it out (indicated by slow "stretching" gestures). 4. OK, you're all right now, its OK (confirmation signal). 5. Keep talking (open/close break movement). 6. We're/you're on time, on the nose. 7. FM TO CAMERA: Are we on time? How is the time going? 8. You have . . . time left (illustrated—2 min and ½ min). 9. Wind up now, conclude action. 10. Cut it, stop, finish, omit rest of item. 11. You are cleared. You are now off camera, and can move or stop action. 12. Speak more loudly. 13. Volume up, louder. 14. Volume down, quieter (sometimes preceded by "Quiet" signal). 15. Quiet, stop applause. 16. Get closer to mike. 17. Come nearer, come downstage. 18. Go further away, go upstage. 19. Stop, keep still. 20. Turn round (in the direction indicated). 21. Move to camera left. 22. Tighten up, get closer together. 23. Open up, move further apart. 24. You're on that camera, play to that camera (sometimes with turning head gesture). 25. Play to the light indicated. (When actors are shadowing, point to light source and to area of face shadowed.) 26. Commercial break (brush right hand over left palm—spreading butter on bread). 27. FM TO AUDIENCE: You can applaud now (may be followed by "Louder" sign). 28. Stop. (For applause, widespread action, etc.). Some use this sign ambiguously: slow to stop; fast to increase. [Reproduced with permission from Gerald Millerson, *The Technique of Television Production,* 11th ed. (Boston: Focal Press, 1985).]

ing control of program length. The cliche "time is money" is all too true in broadcast television. One minute of national advertising time during prime time viewing hours on a major commercial network can cost hundreds of thousands of dollars. The director ensures that there is no loss of time using two methods of timing. The first method, called *front-timing*, is used for the majority of a program. Each major program segment is timed, including the opening up to the first commercial break, the program content up to the next commercial break, the commercial, etc. The director adds up the time of all program segments and the result is the overall timing for the entire program from opening to close. The director uses two timing devices in order to keep track of time. One is similar to a clock; it records total running time of the program. A second timing device is similar to a stopwatch; it keeps track of the length of each particular segment to make sure that the program is on schedule.

The second method used to ensure that a program ends exactly on time is called *back-timing*. The director counts backwards from the close of the show (e.g., 7:30:00 P.M.) to an arbitrary, predetermined point. The switch from front-timing to back-timing allows the director to give easily understood time cues that signal how much time is left before the show goes off the air. The following is an example of how back-timing would be marked on a script for a show scheduled to end at 5:29:30.

5:24:30 5
5:25:30 4
5:26:30 3
5:27:30 2
5:28:30 1
5:29:00 ....30 seconds
5:29:30 ....station break

Remember that directors develop their own unique ways of marking scripts, and this includes time notations. Each director is responsible for explaining the particular symbols to the production staff.

The Telecast

Prior to air time, the director makes a final security check. This includes:

1. Checking props, lights, mikes, graphics, charts, and any other visuals
2. Checking talent for makeup, costumes, ready for positions
3. Checking with audio engineer for all audio effects
4. Checking with video engineer for film, slides, videotape
5. Checking with all additional technical staff (if any) being used

The director usually does all of this final checking with the assistance of the floor manager.

On-air countdowns vary from director to director, but the general approach is the same. Directors notify all talent and staff that there are five minutes to air time, announcing the passing time at one-minute intervals until the final minute. During the final minute prior to air time, most directors announce the thirty-second mark, the fifteen-second mark, and the ten-second mark. At ten seconds, directors add the phrase, "Stand by." Following is a typical countdown by the director; statements in parentheses give you an idea of what actually happens:

> "Five minutes to air—everybody in places." (Director contacts the floor manager via intercom. The floor manager then relays the message to all studio personnel.) "Three minutes to air—stand by." (To floor manager via intercom.) The director then contacts camera operators via intercom and says, "Pictures, please" or "Ready shots." (The camera operators then frame their opening shots.) The floor manager is notified of the two-minute mark by the director via intercom. The floor manager makes a

last minute security check. At one minute the director notifies the floor manager who announces, "Ready and quiet in the studio." The director then issues rapid commands as the zero point approaches. "One minute. Pictures, please. Thirty seconds. Fifteen seconds. Stand by audio theme. Ten seconds. Ready theme. Ready title slide. Hit theme. Up on slide. Theme under. Take 1. Ready 2. Take 2." (At the zero point, the theme music is punched up, as is the title of the show. Theme music is reduced and the director punches up camera 1 on the switcher, replacing the title of the show with the first shot. The director then moves into the second shot.)

For different shows, the actual commands will vary, depending on the materials used (i.e., slides, films, graphics, videotape, etc.). But the wording is similar. "Ready ______. Take __________." The *ready* alerts the particular crew member whose material is going out on line. The *take* command puts it on line. Either the director punches up the switcher on the *take* command or the director has a technical director make the punch on the *take* command.

Nonlive and Remote Productions

The development of videotape greatly changed the nature of broadcast television. Few television shows are now broadcast live; that is, as they happen. The vast majority of television shows are videotaped or filmed for broadcast at a later date. The exceptions are daily news and some daily talk shows, sports telecasts, and some special entertainment programs. A special news crisis will lead to a live, on-the-scene program. Nonetheless, most television directors no longer face the extraordinary pressure of having their work transmitted to the public at the very moment it is being enacted.

Television shows videotaped in studios for broadcast at a later date are directed in a virtually identical manner to everything covered so far in this chapter. Of course, certain differences do appear. The on-air countdown becomes a videotape countdown and the director will say "One minute to tape." With tape, directors often will have the luxury of reshooting a scene for greater perfection. During the videotape editing process, the director can make further adjustments.

One style of direction for television that is quite different from traditional broadcast television direction is *shooting film for broadcast*. Many prime time television programs are first produced on 35 mm film. Directors prefer the greater richness that film images give in comparison with videotape. When film is used as the primary medium, the environment is quite different from the one described in this chapter. There are no switchers, racks of monitors, special effects machines, or audio consoles. There are lights, microphones, audio recording equipment, props and film cameras. Directing for film requires competency in the *cinematic style* of shooting. This style relies on one major camera only and the director must be a virtuoso of composition in order to create spatial sense as the camera moves from position to position. Today, many television directors can shoot film for television and thereby move between the two industries.

A second style of direction for broadcast that varies from studio directing involves *remote* video productions. In remote production there is no television studio or adjoining control room. Instead, there is a mobile control center: a truck or van that approximates a television control room in miniature. The larger and more expensive remote trucks perform most major functions found in studio control rooms. Remote production material can be telecast live or stored on videotape for use at a later time. Remote productions take place away from the major studio so the director's responsibilities alter somewhat, especially during the preproduction phase.

In the preproduction phase, the director must make a preliminary site survey. This survey focuses on security, electrical power, telephone and other communication links, and equipment placement (including the mobile van or vans). Remote productions require an acute safety awareness on the part of the director. Equipment must be secured against weather damage and accidental damage caused by interested members of the public. Production staff must be protected against possible endangerment by large crowds. On the other hand, the public cannot be exposed to risk and injury because of cables, heavy equipment movement, and electrical shock. All these factors must be taken into account by the director.

A crucial remote site survey involves the question of electrical power and communication links. How much electrical power is available and where is it located? Are there back-up systems in case of primary system failure? How will the director communicate with camera operators and other staff who are scattered over a wide area? (For example, if an intercom system requiring cables is used, the production is then exposed to problems of greater injury risk and possible intercom failure due to cable breaks.) Further, how will the director communicate with the station? This includes transmitting the remote production to the station (or other sources) as well as receiving information from the station or other sources (e.g., an audio feed from the station). These communication questions are crucial to the success of a remote production and the director is held accountable for the final result. Experienced directors bring trusty engineers on remote site surveys to assist with these power and communication needs analyses.

Finally, the director must decide on camera placement and movement, lighting equipment and audio locations, and position of the mobile control center. This often requires improvisation because traditional camera placement may not be possible on remote shoots. Physical constraints imposed by crowds, buildings, and access to the scene of primary action may require the director to abandon some traditional camera shots and invent plausible substitutes. Remotes are a challenge even for the experienced director. Young directors are given experience with small remote events and work up towards major events such as sports spectacles or national political conventions.

The beginning TV director must acquire technical competency with a variety of equipment; however, technical competency is only the first step. Technique is used to produce creative images and a successful director must advance beyond technical concerns in order to create artistic images. This is possible only after a great deal of practice with professional television equipment as well as intensive study of visual composition for motion media. A professional television director must know every aspect of the production process. The director is the leader, the center of the television production world. A director who hesitates, who is not calm, who is anxious about the understanding of equipment will never become a successful professional director.

BIBLIOGRAPHY

Armer, Alan A. *Directing Television and Film*. Belmont, CA: Wadsworth, 1986.

Arnheim, Rudolph. *Art and Visual Perception*. Berkeley, CA: University of California Press, 1954.

Burrows, Thomas and Donald Wood. *Television Production: Disciplines and Techniques*. Dubuque, IA: William C. Brown, 1978.

Millerson, Gerald. *The Technique of Television Production*. 11th ed. Boston: Focal Press, 1985.

Millerson, Gerald. *Video Production Handbook*. Boston: Focal Press, 1987.

Oringel, Robert. *Television Operations Handbook*. Boston: Focal Press, 1984.

Rabiger, Michael. *Directing the Documentary*. Boston: Focal Press, 1987.

Verna, Tony. *Live TV: An Inside Look at Directing and Producing*. Boston: Focal Press, 1987.

Wurtzel, Alan. *Television Production*. New York: McGraw-Hill, 1983.

Zettl, Herbert. *Sight, Sound, Motion: Applied Media Aesthetics*. Belmont, CA: Wadsworth, 1973.

Zettl, Herbert. *Television Production Handbook*. Belmont, CA: Wadsworth, 1984.

8

Performing

William Hawes and Robert L. Hilliard

The principal means for the presentation of the writer's ideas, toward which all elements of television production focus, is the performer.

Without effective interpretation on the part of the performer or performers, the best script, the most innovative producing, the most creative use of equipment, the most artistic setting, the most imaginative directing do not reach ultimate fruition. Although the performer the audience sees may not be in human form—animation, electronic effects, puppets, animals, and other means have been most effective in television—this chapter concentrates on the human performer. Included in this category are actors, actresses, announcers, newspersons, singers, dancers, comics, MCs, interviewers—any and all persons who appear professionally on our television screens. Some are household words, stars of network or nationally syndicated programs. Some are known only in regional or local markets, but frequently are just as much celebrities in a particular town or area as a nationally known performer. Some performers are not TV professionals in the usual sense, but are known in other fields. By virtue of their status they appear frequently on television, either on network television (usually in interview shows) or, as is the case with most, as local celebrities on local stations. Such people include sports figures, politicians, citizen advocates, industry and labor leaders, and artists.

Television, like the movies, tends to cast performers by type. Ability frequently is incidental. Does the actor or actress have the right look, the right voice, the right movement for the particular role? Does the news announcer or anchorperson seem as if he or she will elicit a sense of confidence and a high rating from the viewing audience? Most performers start out as actors or actresses. Johnny Carson did not start out to be an interview-variety show host; he is still an actor, but in a nondrama format. Many newscasters, especially those now older, began as newspaper journalists. Both groups gravitated to bread-and-butter jobs in broadcasting, frequently at small radio stations, doing disk jockey, news announcing, and commercial spot work. Some eventually moved into television. The television performer appears in all kinds of formats: drama, news, sports, announcing, commercials, game shows, interviews, panel moderating, music shows, variety programs, children's programs, women's shows, instructional programs, minority and ethnic programs, features, documentaries—every conceivable format in which a live performer may appear. The growth of industrial programs, or corporate video, has opened new opportunities for performers. The rapid expansion of cable television is another growing area for performing jobs.

THE MEDIUM

A distinct opportunity for the good television performer is the chance to use, by the very nature of the medium, a combination of both representational and presentational acting techniques. Even as the close-up permits a delineation of the intimate, subjective thoughts and actions of the character, the opportunity and sometimes the need to work directly to the camera results in the person-to-person presentational effect. On occasion the performer is required to combine both techniques virtually at the same time, a fine means for effective communication to the audience, but not an inconsiderable task.

Concentration is one of the key disciplines of the performer—and probably the most difficult thing to achieve in television. It is not possible to isolate oneself as on the stage and use "inspiration" or to "lose" oneself in the character. Even while having to understand, believe, and feel the character's motivations and actions in the creation of the role, the performer—and this includes non-drama situations, such as news announcing—must be aware of the cameras and the directions, watch the floor manager for cues, look out for the mass of equipment and numerous technicians running back and forth, and constantly be ready to adapt motivation and action to the exigencies of a technical medium. The television performer must develop intense concentration and full awareness of the outside surroundings at one and the same time. The problem is to remain calm and collected—and, sometimes, sane—in the midst of the confusion and mechanism—and madhouse—of TV.

The Studio

The performer works in an environment of cameras, microphones, lights, other technological equipment, and innumerable technicians and performers. The performer must adapt to the technical requirements of the electronic medium. For example, one must be aware of the placement of the microphones and of the carefully lighted areas designated on the studio floor in chalk or masking tape: the *marks*. Being off one's mark, even slightly, may sometimes make a substantial difference in the light level or composition of the shot. Studios that do not have enough sound equipment or lighting instruments to cover the entire floor create electronic problems: a performer may come out in deep shadow or a distorted color, or may sound as if he or she is speaking from another part of the scene. Unless the electronic results from the cameras are identical, a performer may look good on one camera and poor on another.

The quality of the studio and the amount and state of the equipment available frequently determine whether a program is to be taped, filmed, or, in some smaller studios, done live.

Makeup

Beyond the type-casting approach, in which many producers and directors select performers on the basis of how closely their physical appearance approximates that of the character or job-role being auditioned, makeup is an important factor in achieving the appearance desired on the TV screen. The performer must develop on the outside, particularly through the face, that which expresses the inside: the wise, knowledgeable, and serene national newscaster; the open, friendly, and honest commercial announcer; the rugged and hardheaded "Mr. T."; the good-hearted, unsophisticated, tolerant "Alice."

Facial makeup is somewhat like the lights and setting of the stage: highlights, shadows, lines, and angles. The key is in application. Some performers, because their role-personalities call for it, use heavy makeup—Liza

Minnelli and Boy George, for example. Others, such as newscasters, use comparatively little. For most TV performers, most of the time, makeup is used to give the skin a healthier tone and smoother texture under studio lights.

A water-base solid, cream stick, or grease paint blends the skin color into the hairline, back to the ears, and slightly under the chin. Light-complexioned men frequently use a tan or rose hue one or two shades darker than their natural skin. Light-complexioned women frequently use a pink hue. For the men in one daily soap opera a foundation of Sun-glo by Clinic is applied to the face and Light Egyptian by Max Factor to the body. For women, makeup for the face varies greatly, but Tan II by Max Factor is commonly applied to the body. Black and dark brown-complexioned performers usually use only a powder to reduce any facial gloss to a matte. Scars and blemishes are hidden by the application of a water base foundation that matches the skin. Caucasian performers who seek a medium gray-scale tan may sit in the sun or under a sunlamp to obtain it. There is no ideal answer for makeup; each performer has different requirements, depending upon that person's natural skin, lips, and hair coloring, and the needs of the TV role. Many performers prefer to apply makeup with a brush, to protect the skin tissues from rubbing, while others blend in makeup with their fingers.

As noted earlier, the performer must be aware of lighting patterns and *marks*. That is because lighting and, in many instances, camera angles and distances affect appearance as much as does makeup. Even a foot away from a performer's mark the light level may drop 50%. For performers, especially with a dark complexion, the result is low visibility. For dark-skinned performers proper lighting contributes more to a three-dimensional appearance than makeup. The performer's appearance relies, therefore, not only on a makeup expert and the performer's own knowledge and skill, but on many technicians involved in the entire production.

Voice

There are many regional dialects in the U.S. and, according to many specialists, three major speech types: New England, southern, and midwestern. Midwestern speech is generally accepted as the most desirable for national performers, based on the theory that no member of the audience should have any difficulty in understanding everything the performer says. Local and regional stations, however, may prefer local and regional dialects, to establish closer identification with the audience and more credibility in announcing commercials. It has been said that President Lyndon B. Johnson could speak with or without his Texas twang, depending upon the audience he was addressing. A growing school of thought is that multidialects on TV lend variety and richness to the language and speech patterns of the country. Unlike his immediate predecessors—Presidents John F. Kennedy (New England), Lyndon B. Johnson (southwest), Gerald Ford (midwest), and Jimmy Carter (south)—President Ronald Reagan offered an excellent example of general American speech; that is, his voice had no regional or ethnic intonation. Indeed, that was the kind of speech he used in most of his roles in his previous profession of media performer.

Some performers have speech styles resulting from a functional or organic defect; one network newsanchor has a faulty or "dark" l, another has a slight lisp. Such defects may be the result of psychological causes, functional causes such as improper use of the tongue, or organic causes such as missing teeth or improper overbite. Obviously, one can reach the top of the profession with such a defect. But young performers should be aware that it is easier without one. A speech psychologist, voice coach, or speech pathologist should be consulted if a defect is suspected. Many beginners do not realize that they may have a speech defect or severely pronounced regional dialect. Dialects are preserved by the speech patterns of teachers and friends in the beginners' home areas. Some young performers may even work successfully in TV in their regions, but unless there is something special about their personalities that obviates a well defined speech localism, moving to national network performing will be difficult. A Cotton Bowl princess was once told by a producer of TV commercials, "You're beautiful and photogenic, but we can't use you until you learn how to talk."

Awareness of the defect or dialect is the first step. Hearing the desirable model after correcting any psychological, functional, or organic problem is the second step. Third is repeating the model at the conscious level until the difference is learned. The final step is to allow the desired speech pattern to sink into the unconscious level so that during the pressure of an on-the-spot newscast, controversial interview, or dramatic scene the performer does not revert to a prior speech mode or dialect. The procedure may take a long time, and there is no easy substitute for the hard work required with specialists, listening to models, and hours of self-evaluation with a tape recorder.

Movement

Because television is a close-up medium, movement is minimal. That is, facial expressions must be carefully controlled, hand, arm, and body movements should not be broad and sweeping, unless specifically called for by a director for a long shot, and walking or running has to be oriented to the movement of the camera and the size of the TV screen.

Any movement that is not absolutely essential to convey the idea, feeling, or message should be eliminated. Unnecessary movements distract the audience. Except for acting situations where they fit the character, movements such as finger-tapping, ring-twisting, eyebrow-raising, ear-pulling, and squinting should be avoided. That does not mean that a performer should be stiff and stilted. The key is to convey a sense of comfortableness, relaxation, and trust to the viewer. One should practice being graceful and fluid, with good posture. For nonacting performances, remember that there is a back to the chair. Sink back a trifle, combining a sense of relaxation while retaining high levels of energy and stature.

Studying theatre movement, dance, and gymnastics can be very helpful. Remember that although some performers are presenting their own personalities to the audience, most performers are interpreters, presenting the purpose of the writer, with mental attitude and movement oriented toward that goal.

PERFORMANCE TYPES

All performers, including entertainers, journalists, and announcers, combine a certain amount of acting ability and technique with their special expertise. Even performers who ostensibly are portraying themselves are creating roles. The host of the talk show may personally and privately not be outgoing, warm, witty, and charming; the stand-up comedian may in reality be depressed and introspective; the announcer of commercials may spend long hours practicing how to convey a sense of personal trust and wisdom to the audience; the newscaster has to develop a style that reflects what the audience most respects and appreciates in a news reporter or commentator.

The principles of acting provide a base for all television performances. The basic material presented in the section below on acting applies to all performers; the sections following it reflect specific considerations relating to nondramatic types of television performance.

Basic Acting Techniques

The best training for the television performer is the stage. The basic elements of acting are learned in the theatre and if any talent does exist, it is there that it has a chance to come to the fore. (It is a peculiarity of film technique that although there are many fine performers in that medium, the ability of the medium to edit, create, and recreate both visual and sound aspects of any performance permits a person of virtually no acting talent to become a successful star.) The stage performer will find that his or her abilities are transferred to television, insofar as character interpretation and development are concerned, almost in toto. The special characteristics of the medium, however, require special adaptations for most effective use of its potentials and most efficient adjustment to its restrictions.

For the theatre-trained performer, television lacks most importantly an interaction with the audience. The performer cannot feel the audience response and play the subtleties of character delineation accordingly. Everything must be fixed beforehand, in anticipation of what the reaction of the small group watching in front of any given television set is likely to be.

Filmed or taped (with film-technique) television production—that is, where the scenes are shot individually, usually out of sequence—prevents the performer from building or maintaining a rising intensity of continuity of character development. On the other hand, live-type taped television—that is, recording in script sequence—does permit this and, in addition, provides the performer with the continuity of stage acting and the close-up intimacy of the film at the same time. Even here, however, the performer's advantages are complicated by restrictions. Because the sustained performance is in a close-up medium, the performer must never for a second step out of character. The performer must be acting or reacting to the other performers every moment, not knowing when the camera may pick him or her up—even when supposedly off camera, but still on the set. The performer must "freeze" in character before the next shot in case the camera comes in a second earlier. We have all watched the embarrassment of a performer who suddenly leaps into action after the camera has unexpectedly come on, particularly on live news or interview shows. The television performer must always be in the scene, listening, feeling and conveying the feeling.

The technical needs of television and the necessity of the director to have every shot planned clearly beforehand, with no unwarranted deviations once on the air, require the performer to make frequent compromises in motivations, interpretations, and movements. For example, to meet the requirements of a particular shot, the performer may have to make a movement that is not clearly a part of the character being portrayed. Accordingly, that performer may then have to adjust character interpretation to find valid motivation for that particular movement. Usually, this kind of adjustment is made during the rehearsal period. Sometimes it is more difficult. For example, in one television production the actress was supposed to jump out of the armchair in which she was seated and cross the room to her injured husband. The camera, however, instead of getting the previous close-up on her from a distance with a long lens, as planned, had dollied in with a short lens and was completely blocking her way out of the chair. She had to change her character's motivation and action for that moment to justify her staying in the chair until the camera could be pulled back, and an alternate shot pattern developed by the director. With continuous-action live-type taped productions in front of an audience, this could be a problem. With standard filming or taping, the sequence would be replanned and reshot.

Because of the impact of television's special characteristics, it is important that the performer become acquainted not only with the director's approaches and problems (see Chapter 7), but with the technical needs and potentials of camera movement, switcher-fader transitions, lighting, sound, and special effects (see Chapter 5). A familiarity with the terminology and uses of equipment and directorial techniques can be of immeasurable help.

Pointers for Television Acting

The following are some of the more pertinent considerations for the television performer in making the most of the medium's potentials and at the same time adjusting to its restrictions. The first five items are especially basic in their applicability to all types of television performing.

1. Television, as a close-up medium, requires movement, gesture and expression to be both natural and restrained at the same time. The performer must use the face and body with the fullest control, scaling down the entire pattern of movements and conveying all things with the minimum specific motion possible—unless told otherwise by the director. An economy of movement is important and each cross, gesture, or move should be purposeful and the performer should clearly validate the reason for it, making every action count. Excess movements are especially distracting because they are so close to the audience. The slightest facial expression often can serve for what would have to be a gross movement or gesture on the stage. A facial expression that conveys something at all times is important, as is an avoidance of grimacing. Arm and hand movements must be carefully controlled or the gesture may go right off the edge of the television screen. Nose-to-nose playing may be required in two-shots of what may otherwise be natural conversation between two people; on the television screen the distance will look normal. Because of close-ups, not only must the performer avoid artificial and exaggerated movements, but must have an ease, a grace, and a naturalness, as opposed to giving the audience the feeling that he or she may be ill at ease, awkward, or self-conscious. Even static pictures can be given the essence of movement sometimes by the change of camera angle or distance.

2. A performer must be aware of the mike placement and pickup patterns to help maintain the proper level of vocal reception while retaining a well modulated tone. For example, in moving across a set a performer not only has to walk to the pattern set up for the camera, but must time and space the vocal delivery to fit the microphone pickup pattern.

3. Although a performer must always be aware of the location of the camera (the tally light on a camera indicates that it is the one that is on), it is important to never look directly into it unless so ordered by the director, or by the form of the program. Most TV dramas are representational in style; many commercials, the newscast, and the variety program, among other types, are presentational in nature and demand direct rapport with the cameras. In some instances in the representational drama the performer is required to "cheat"; that is, for the purpose of an effective close shot, to turn somewhat away from the other character and toward the camera; the performer still seems to be in direct relationship with the other character, but permits the camera to pick up the desired shot of the face or body.

4. Because of the comparatively little rehearsal time in the pressure- and budget-controlled commercial television field, the performer must memorize quickly and adhere to what has been agreed upon.

5. A performer must learn to take cues from the floor manager without having direct eye contact and without being distracted from the business under way.

The following items apply more specifically to the dramatic performance situation.

6. The professional performer is constantly writing, making clear notes of all the directions given by the director and making certain that they can be executed with precise detail.

7. Precision is important. Because a specific shot has been prepared by the director, the performer must be able to repeat on cue the exact spatial position and bodily relationships set up during rehearsals. All directions must be memorized as accurately and fully as the character's lines.

8. A performer must learn to "hit the chalk mark," the term applied to the exact place the performer has been rehearsed to be in for any given shot. The slightest error may throw off the entire composition of the shot. Yet, the actor or actress must hit the mark without being obvious or mechanical about it.

9. A performer must never drop out of character.

10. Unlike the theatre, where, after the curtain goes up, the director is gone, in TV the director is still there, calling the shots. Since the director controls the cameras, the performer must follow directions explicitly. The performer must voice any complaints with any direction or acting requirement during an early rehearsal before the pattern of the show is set.

11. Although performers in commercial television usually do not have to oversee their own costumes or apply their own makeup, good performers always work as closely as possible with the costumer to see that their costumes fit the needs of the characterization, and know enough about makeup to clarify the characters' needs to the makeup person or to put on their own, if necessary. (Many experienced performers insist on applying their own makeup.) As noted earlier, makeup for television must be extremely light and subtle. One must be careful of aging during a show, of using beards and wigs, and of applying lines to the face. Usually, only a lightly applied base with light and shadow tones, as opposed to lining, is effective. One should avoid extremes of black and white in costumes. It is best to stick to the middle tones and avoid clashing colors, and striped or wavy lines.

Announcing

The television announcer needs the desirable vocal qualities of the radio announcer: control of volume and pitch, pleasant quality, clear diction, and accurate pronunciation. The addition of sight, however, necessitates special adaptation. The television announcer's vocal qualities should match his or her physical appearance—an appropriate, optimum pitch, for example. The announcer must know not only the proper use of the microphone, but also the proper use of the camera. More than any other type of performing, this requires the presentational approach—talking directly to the audience, coming directly as a visitor into the home.

In addition to knowing how to "read" well, that is, to present the ideas smoothly and with effective interpretation, a television announcer must have a broad background. This background should include a knowledge of the announcer's duties, limitations, and impact on the audience as a part of the television process; function in relation to the producer and director and to the artistic and routine processes of production; place in terms of the technical and staging potentials and needs of the program; and familiarity with the various commercial announcements, forms of presentation, and program types. In addition, the announcer should have a broad background in the philosophical, historical, social, economic, and political influences of our time.

Above all, the announcer must remember that he or she is a guest in someone's home, not portraying a fictional character. An announcer should behave with the kind of honesty and sincerity that is expected from a guest. Because an announcer is persuading people to do or believe something, an announcer should speak and behave with good taste. An announcer should be aware that no matter how good and effective the vocal presentation, the slightest disconcerting facial or physical movement, or the smallest note of falsity in a gesture may obviate every word.

The announcer conveys not only ideas, but feelings and personality. In that respect the announcer is also an actor or actress, and should be thoroughly familiar with the special requirements of television acting.

Newscasters

The media have gone through periods in which the news was presented by people who may or may not have been journalists, but were attractive personalities. These "readers" do not gather or edit news, but deliver a prepared script. In the early days of radio, newspaper and wire service journalists brought their investigative and reporting abilities to the sound medium, but the visual appeal of television tended to concentrate on personalities who drew larger audiences than the news itself. Some national newsanchors, who have excellent reporting credentials in their own right, such as Tom Brokaw, Peter Jennings, Dan Rather, and Connie Chung, share equally with movie stars as the dominant national personalities. News specialists such as Ted Koppel and Barbara Walters are as popular and well known as any celebrity in the country.

Some news formats de-emphasize the newscaster by concentrating on film or tape of the actual event, featuring the participants in that happening, with an unseen announcer doing voice-over narration. Some news programs have concentrated on newscasters who are strong as journalists, with the attractiveness of their voice, appearance, and manner secondary. Ideally, of course, the most desirable newscaster is one who is a good reporter, editor, *and* personality.

Specific backgrounds and kinds of personalities differ with the kind of news presented: hard or straight news, interviews, sports, weather, soft news or human interest features, among others.

Becoming an information specialist, that is, going into news, is one track to media success for a performer. But it has its hazards. A person in news specializes in reporting natural and human events that dominantly occurr to other people. A reporter witnesses the changes in history and the world from as close a vantage point as possible. Serving as liaison between the event and the media public is an enormous responsibility that requires instant mobility, wide knowledge, deep insight, and an eagerness to find out and to tell the viewers about an incident, accurately, by means of sounds and pictures. The news reporter's eyes and ears must be able to display, often recreate, the essence of an event through precise, computerized editing with such skill that the viewer is never bored, but is always fascinated. This is exciting, fast paced work with complex challenges.

The newscaster must be intellectually prepared not only to report, but to understand and interpret the news. Most stations prefer to hire news personnel who have college degrees in history, political science, or economics and who are in other ways additionally sensitive to and knowledgeable about the affairs of the world.

Entertainers

Singers, dancers, and other variety artists usually have developed their particular acts in night clubs, and adapt them to television. Some have created new personalities primarily effective for television and have developed reputations through TV far exceeding their previous work. A few performers have become fixtures on the TV screen. Some older, established personalities, as well as some rock stars, are in great demand and require large fees for appearing on TV specials or as guests on regular variety programs.

By contrast, the life of an entertainer, even for one who has had good training in acting, dance, music and other arts since childhood, is at best difficult. Assuming the entertainer has developed a talent to a high degree, producers/financiers must be convinced they will make money off the entertainer, preferably more money than they would from investing in other businesses. Individuals who demand high fees for performing become businesses with managers, accountants, lawyers, coaches, and an entourage of advisers. An actor or actress must be able to unequivocally do whatever is necessary to gain power, prestige, and the edge on getting the relatively few jobs available.

The demands on an entertainer are enormous because the entertainer is the entity. An entertainer must train like an athlete by working out in a health club or studio every day so that every mental and physical ability, obvious and latent, is developed. A person's talent in conceptualizing and projecting an image in audiovisual terms is nontransferable, although skillful coaches can provide valuable assistance. The most gifted artists may be obsessive, frustrated, egocentric, and dreadfully insecure. A continuous gag repeated by actors employed at TV studios is: "Have they fired me yet? I haven't read next week's script."

While words of discouragement abound, and data galore can be cited about the slim chances of getting into entertainment successfully, the bottom line to any well trained hopeful should be "Go for it!" The associations, the richness in experience, if not in money, and the wonder of being in a profession one loves make the effort worthwhile.

Personalities

The performers who host talk shows, quiz and audience participation programs, interview programs, panel shows, and similar programs are attractive, frequently unique personalities. Performers such as Phil Donahue, Oprah Winfrey, Merv Griffin, Jane Pauley, Bryant Gumble, Joan Rivers, David Letterman, Mike Wallace, Barbara Walters, Ted Koppel, and others already mentioned come across as strongly as the people who perform on their programs or whom they interview. They specialize in being themselves, or, rather, the public image of themselves that they have developed for the television audience. Even those who may not have been trained as journalists may broadcast the news; even if not trained as entertainers, they may sing, dance, and tell jokes; even if not trained as political scientists, they may interview politicians and other public figures. As with newscasters, the ideal personality is the one who has training and/or ability in particular subject fields as well as appealing to the audience through individual attractiveness. Some of the more successful ones have this combination.

Their strongest ability is to bring out the feelings and ideas and personalities of their guests. They are good catalysts and are able to keep a program running fast and smooth and to make it exciting. The audience finds them fun and/or stimulating to be with. They are usually well versed on a great many subjects. They are able to improvise in extemporaneous situations. These are the superstars of television and the ones the audience usually feels closest to as individual human beings.

Nonprofessionals

With television providing a forum, principally through talk and interview shows, for experts and human interest experiences from all fields, more and more nonprofessional performers who are professionals in other areas are becoming media performers. Politicians, executives, physicians, educators, consumer advocates, and clergy are among those who are becoming, for large and small audiences, TV stars. One dramatic example is Chrysler president Lee Iaccoca. His appearances in his company's commercials in 1986 helped stimulate talk about him as a

U.S. presidential candidate in 1988. The growth of cable has greatly increased the appearances of nonmedia people on the local level.

Many representatives of various professions are successful in person or on speaking platforms, but are not able to carry a favorable public relations image to television. Some professions provide the media with entertainment as well as personalities. The clergy have been particularly successful in this regard, with ministers as dramatic as any actor and with religious presentations approaching the level of entertainment spectaculars.

In the corporate area, the use of closed-circuit television, cable, microwave, satellites, tapes, and other means of transmission for teleconferences, sales presentations, information transfer, training, and other requirements have made performers of many top executives and even mid-level managers.

For years educators have preferred to believe that instructional television (ITV) programs should be taught by teachers. Unfortunately, not all—or, indeed, many—teachers on TV are also performers, and we find many educational television programs dull and boring. Coupled with unimaginative production aimed at merely duplicating on TV the kinds of materials found in the classroom, too many ITV programs are of the "talking head" variety. More and more, ITV producers are becoming aware that TV teachers must be attractive performers. With the proper materials and a script, the good performer, even knowing nothing about a given subject, can make a more effective impact on the student than can the knowledgeable teacher who cannot perform. As instructional TV programs become more sophisticated and take advantage of the medium itself, most important are the teacher-performers who can interact with the student-viewers in direct or indirect participation in discussions, projects, and other means of personal involvement. As in other performance types, ITV needs people with a combination of educational expertise and personal charisma.

No matter what the performance type, however, the goal is to bring credibility to whatever message is presented. Hopefully, the performer not only will bring attractiveness to the presentation, but judgment, responsibility, fairness, honesty, and perspective as well.

PROCEDURE: REHEARSAL AND PERFORMANCE

Some stars have such strong egos or confidence or both that they sometimes think they do not need much rehearsal time. They forget that it was long preparation and practice that made them stars in the first place. (There are a few stars who are overnight fads and whose success is based on a quirk of audience need and public relations; because they may not have any talent to sharpen, they do not need rehearsals in terms of their personal art.)

Television is a group effort. There are not only a great many performers in any given show, but a large technical crew. All of these people need adequate rehearsal if the program is not to be in chaos by the time it is filmed or taped for air presentation. Even stars who do not want to rehearse should remember that the other people associated with the show need rehearsal and they cannot rehearse unless all the elements in the show, including the star, are there.

Types of Rehearsals

Dramas, variety programs, and shows with detailed and complicated segments and complete scripts need complete rehearsals. The drama, for example, may begin with script interpretation, with the director and all the performers present, then character analysis and development, followed by blocking, with camera and other technical personnel observing, and gradually moving toward run-throughs with costumes, makeup and all the technical effects, by which time all personnel associated with the program will be involved. The final rehearsals are on camera, in the studio, ironing out all problems before taping or filming or, in some small, local or public stations or cable systems, before a live performance.

Rehearsal hours are limited by union contract and are expensive, and it is critical that all performers (and other personnel) be precisely on time.

Some programs, such as panel shows, audience participation programs, interviews, and similar productions that do not have complete scripts and include some extemporaneous (although carefully planned) sequences, have what are called abbreviated rehearsals. This includes the opening of the show, transitions between program segments, and the closing of the program. Occasionally, in small stations without adequate personnel and on programs involving nonprofessionals from the community, there is no time for rehearsal. The station announcer assigned to host such a program must simply practice the opening, transitions, and closing alone, and communicate some basic instructions to the nonprofessionals when they arrive at the studio.

With the film or tape recording and editing techniques of a given program, some performers rehearse and perform only those parts of the script in which they appear, and sometimes do not see the entire production until they watch it on television with the rest of the audience. Some programs are recorded on tape in continuous-action sequence, sometimes in front of live audiences. Here the performances are akin to the continuing concentration required for a stage play. Audience participation shows may be shot in multiples, that is, several programs in one day in front of different audiences.

The performer should be so well prepared, thoroughly rehearsed, and confident in the role or job that it is second-nature to be flexible and to adapt to any form of final performance and to any emergency.

Cueing

When a performer arrives at a studio for a filming or taping, preparation should be complete. The script has been memorized or is on cue cards. The blocking has been rehearsed or is generally understood. If there have been sufficient rehearsals, the performer is already acquainted with the performing area and with costumes and makeup. Usually, a final technical rehearsal precedes the taping or filming. Once the performance is under way, the director is in the control room, separated from the performers. Yet, there is a continuing need to communicate. This is where cues come in. During the rehearsal period and before the airing or taping, the performer has been given visual and verbal cues or check-points. These continue during the performance.

Oral and visual cues are given by the director through someone on the floor, usually the floor manager or the camera operators, who have intercoms linking them with the control room. A performer must know who is giving the cues, where that person will be, and how visible the cue will be. Once cued, the performer must react instantly. In stop-and-go taping and filming situations the director halts the action after each shot and gives directions through a loudspeaker or meets directly with the performers. In the continuing-action taping approach, the director may stop and repeat the action, but usually gives the new directions from the control room.

Typical oral cueing prior to the beginning of a TV program is announced by the floor manager: "One minute to air (or tape) . . . thirty seconds. Quiet in the studio . . . Ten seconds. Stand by." (A countdown may begin: "Five, four, three, two, one—") "Cue."

At the ten-second "stand by," the floor manager raises a hand with the flat of the palm toward the talent or points an index finger toward the on-air lens. On cue from the director, received via a headset, the floor director points directly at the performer, who responds instantly. Most visual cues concern timing—the minutes or seconds remaining before the program is over, wrap up, stretch, or speed up. Some cues deal with movement—the floor director waves the performer from one camera to the on-air lens of another, especially if the talent cannot tell from the tally lights which camera is on the air. A performer may cue the director by placing the palms of the hands on a desk in front, or on top of his or her thighs to indicate that he or she intends to stand. Some cues have to do with transitions: if a floor director taps the palm of one hand with the index finger of the other, this signals the talent to go to a spot announcement or commercial; if the floor director assumes the pose of looking through a lens while cranking a camera, this means go to film. For the most part, as long as the performer remains on his or her marks, sticks to the script or program outline, and performs as the program was rehearsed, cueing is minimal.

At the show's conclusion, the talent does not move until the floor director gives the cue, such as "That's it. Good job. Thanks a lot," or until the cameras have obviously broken to another set. A common error for a novice is to move, especially to take one's eyes from the camera lens, or grimace prematurely. Such action might ruin the take.

The Camera

The camera shows everything in its field of view. Whether a performer is scratching a leg or adjusting a microphone or making an unpleasant expression, the camera will show it. Performers must assume that they are on the air all of the time during the airing or taping of a program; therefore, they never do or say anything that could not go over the air.

Some studio cameras are frequently big and clumsy. They may move slowly from set to set. The zoom lens adds fast and fluid variation from a wide shot covering the entire set to an extreme close-up of the performer. Minicams move quickly in and out of studio situations. No matter how many lenses are visible, a camera has only one that is "hot" or on the air. This hot lens is in a different position on different cameras depending on the model. The performer should know, from rehearsals or briefings, which lens is hot. Each camera has a red tally light indicating when it is on the air.

Generally speaking, except in a dramatic program or an interview, a performer looks directly into the hot lens when he or she is on. Performers relating directly to the audience (such as singers, newscasters, and speechmakers) frequently perform for a mythical person fixed somewhere in the dark space above the TV camera. Frequently, performers who must look directly into the lens transfer their gaze from one camera to another as the shots are changed. Such movement should be signaled in advance by the floor director by waving a hand from one hot lens to another. Prompting devices—scripts, cue cards, and electronic devices—are commonplace and performers should get used to working skillfully with them.

The Microphone

The engineer will adjust the level of the various microphones on the console so that every performer's voice blends in with the rest of the program. It is important to cooperate with each request for a "level check, please." There are some general techniques performers should know for the most effective use of microphones.

If one must have a microphone around one's neck, it should be placed under the clothing—a tie or scarf will do. If the clothing muffles the sound, the engineer will say so. The lavalier must be fastened. If it moves, it may hit buttons or jewelry; even swinging over the surface of a shirt may be noisy. A jacket or patterned blouse helps women, particularly, to draw attention away from the microphone. To minimize its visibility, the mike may be attached to one's belt, provided enough slack is left so that it will not pull on one's clothing. If one must hold a microphone while standing, it is important to remember that the long cable will show. It is distracting to see a performer wrestling with a long microphone cable or being trailed by a long black cable resembling a rat's tail. It is best to preplan to have ample cable during the performance.

If working with a microphone stand, one must remember that it causes a resounding thud if accidentally kicked. If the microphone is suspended on a boom or is overhead, one must avoid the temptation to move out from under the beam.

In general, while singing or speaking, one microphone, preferably unseen, should be enough. For round-table discussions, one good omnidirectional microphone is adequate. For panelists seated in a straight line, usually one microphone for every two or three people is satisfactory. One should not attempt to pass the microphone, because it will make noise. All principal singers and instruments in a musical group must be miked separately. An orchestra, however, may be picked up on one microphone suspended in the center of the auditorium, in addition to personal microphones for selected participants.

UNIONS

Networks, recognized film companies, larger stations, and, in some parts of the country, smaller ones, are generally unionized. Some sections of the country that have strong antiunion traditions or laws do not have performers' unions in broadcasting. The nation's fifth largest city, Houston, for example, is primarily a nonunion market. Nonunion talent get whatever compensation the highly competitive performers' situation will allow. Union performers get no less than the minimum agreed to by union-station contract, and this is virtually always higher than nonunion fees.

Many performers belong to more than one union. The American Federation of Television and Radio Artists (AFTRA) and the Screen Actors Guild (SAG) represent media performers. Actor's Equity Association (AEA) represents performers on the legitimate stage, in stock companies, and in industrial shows; the American Guild of Variety Artists (AGVA) covers nightclub performers. AEA and AGVA members who perform on television join AFTRA and/or SAG, as appropriate.

SAG, originally the theatrical motion picture performers union, retains jurisdiction in that category and represents performers in all TV shows and commercials made on film. AFTRA covers the nonfilm categories, principally in news and other nondrama areas. AFTRA and SAG contracts establish minimum wages called *scale*. Scale is tied to the lengths of the roles and programs and to usage, whether local, regional, national, or international use. Contracts also cover residuals for re-use of programs. There are different scales for principal performers, for those speaking under five lines, for nonspeaking roles, for extras, for specialty acts, for voice-over, for dancers, singers, groups, and the various other performer categories. AFTRA and SAG contracts define rehearsal hours and days, guaranteed days of employment, meal and rest periods, credits, wardrobe, hair and makeup requirements, retakes, understudies, stand-ins, vacations, holidays, overtime, remotes, auditions, travel requirements, dressing rooms, and other working conditions. AFTRA and SAG have won for their members a pension and welfare plan and a health and insurance plan, both paid for by the producers.

Initiation fees and dues for AFTRA and SAG vary in different sections of the country. Combined initiation costs in New York and Los Angeles, for example, are about one thousand dollars, while in a smaller market they might be only a few hundred dollars. Dues are assessed on a sliding scale, depending on how much the member has earned during the year. The more a performer makes, the higher the dues. Performers cannot become union members until they have been offered and accepted a contract to perform in a program or at a station covered by union contract. Typically, a nonunion performer may be covered on that first job by the Taft-Hartley Act, which allows a nonunion performer to appear on an otherwise union project because he or she is deemed the best available person. This job also serves as a minimum professional requirement for entry into the unions. Membership, therefore, is limited to bona fide professionals.

A PERFORMANCE CAREER

Every performer has three primary concerns: "How do I look? How do I sound? How good is my material?" Good material can make a performer. Conversely, an exciting personality can enhance mediocre material. Because television is essentially a "personality" business, the performer is usually more marketable than the material. The person—that is, the sound, the look, the image—can be packaged and delivered to an audience over and over again until the right wrapping has been found. If the audience likes the performer's image and sound, that performer might have instant and overwhelming fame through a single TV appearance. Such situations, despite the glamour stressed in fan magazines, are extremely rare. Most performers who make it have done so through years and years of study, practice, struggle, and failure. Because the sacrifice is great in time and money, comparatively few persons are willing to really prepare themselves for performer roles. That is one reason why the few who do are sought after and highly paid.

If a would-be performer has a unique look, a unique voice, some unique talent, the capacity for hard work, sufficient ambition to prevent him or her from being easily discouraged, and great patience, then it's worth a try. And a performer must also have foresight. One must be able to recognize an opportunity when it comes along and drop everything else to take it. And then one must work inexhaustibly, often without reward, often with criticism, to remain in competition for professional status.

There is no certain or single way to become a successful performer in television. One can go to school, take courses in communications, speech, drama, journalism, music, art and other subjects. One can be on the debate team, have leads in plays and achieve recognition in related fields. One can go to college, vocational school, professional schools, and universities. One can gain professional experience in summer stock, nightclubs, and radio and TV stations. All of these can be helpful and a performer can do them all and still not make it.

Nevertheless, in recent years more and more colleges and universities, in association with the television, cable and corporate video industries, have established internships that enable the most promising students to make a transition from the classroom to the marketplace. After graduation the small to medium cities have sought young talent that brings freshness to the TV station and daily experience to the performer. Two or three years of professional experience, preferably under the guidance of mature working professionals, in addition to a good education and talent, opens the doors to larger markets. Entertainers head for entertainment centers in Hollywood and New York City, where they can anticipate further study, working part-time in undesirable jobs, making agent-producer contacts, *showcasing* in all media, and generally learning how media work. Some can judge the status of their talents and their realistic opportunities for success after a few years. Some never do and go for decades vainly reaching for a star.

Performing is unstable. Only a small percentage of professional performers—those in the professional unions by virtue of already having worked professionally—have jobs at any given time. Most performers pay the rent by working as salespersons, in restaurants, as stock clerks, in secretarial positions, and in other jobs that provide them a living but are easy to leave at a moment's notice in case the break comes. Some highly talented performers spend twenty or more years of their lives waiting on tables, never or hardly ever working in their profession, in television. It is estimated that some 80% of AFTRA and SAG members, the bona fide professionals, make less than $5,000 a year in their chosen professions, as performers. Yet there is the handful who make it big, and for most young people who wish to be performers, the dream is always there.

To assist in making that dream a reality, the industry requires a resume of one's background (about a typed page), a well produced videotape or excerpts of one's experience, and a list of credits, paid and unpaid, that show media appearances. Photographs, 8″ x 10″, overprinted with vital statistics (hair color, age, height, weight) are essential for many jobs in entertainment. An extensive photo collection, known as a *book*, is required primarily of models and most actors and actresses.

Salaries vary greatly from $1 million plus a year for a superstar newscaster and interviewer on a national network to $12,000 a year for a combined newscaster, interviewer, board operator, and time salesperson in a small market station. Newspersons in middle market America have salaries in the $18,000 to $35,000 range. Larger cities pay $50,000 and up for news anchorpersons. The largest markets put some newscasters in the $100,000 and up category. If a TV station invests in promoting a talent (billboards, news stories, cross-plugs), it considers that talent a money-maker and will pay accordingly. Many dramatic and entertainment stars earn in the hundreds of thousands and millions and many who are not so well known but who work regularly earn very large salaries. But it is important to remember that competition is intense and that success is the exception rather than the rule.

Locally originated programming is primarily of the news and information genre. It is relatively stable and good income employment. Local entertainment is produced primarily during times of high station income and is therefore unstable, infrequent, and low paid. Major programs, including nationally distributed shows and most regionally distributed shows, whether for network or syndication use, are produced in Los Angeles or New York. Commercials are produced in additional places, such as Chicago, Cincinnati, Dallas, and Atlanta. Local commercials are usually produced locally. Performers interested in free-lance performing, as opposed to staff jobs with stations, will have to establish themselves in those cities where the kinds of programs they might be on are produced. Most performers now live in either Los Angeles or New York, with the successful ones maintaining residences in both places.

A performer must assess himself or herself and ask, "What is there within me that can make a unique contribution to television?" Introspection to find the nature of one's own being is extremely difficult. It can be painful and contrary to peer group identification and acceptance. It can be inconsistent with one's personal and private psyche. A TV performer who makes it is, at least in public, different, and must accept that special place to maintain recognition and success.

SUGGESTIONS FOR BEGINNING PERFORMERS

Ambition. Most successful performers are highly aggressive. The industry interprets anything less as not wanting the work badly enough.

Appropriate Material. Performers should select material that they know they can do well. News departments as well as dramatic programs type-cast. Performers should develop material that will help them to sell themselves; that is, a basic marketable repertoire.

Charisma. Performers appear bigger than life. Media tend to filter out their subtle qualities, and so performers must exude high energy on camera. They must have (or seem to have) fun at what they are doing. They must create the impression that they are a joy to be with. They smile a lot. A smile is perhaps the most valuable physical attribute a performer has. They will go to great lengths to get an audience to look at them, to want to be with them, to "love" them.

Concept. Performers should have ideas that show fresh insight for news stories, features, demonstrations, reports, or entertainment segments.

Contacts. Performers should not fail to develop contacts at hand before expanding to distant locations. The base for interviews, prestige, and opportunity begins at home.

Image. Performers should strive to be like the successful role models of the day, but unique at the same time. A performer should appear slightly different, memorable.

Language Skills. Performers must use correct grammar and have good English language skills. Errors show up in extemporaneous situations such as talk shows, telethons, and interviews.

Mechanical Skills. Performers must develop necessary ancillary skills that give them some degree of control over their own product, such as an ability to edit audio and videotape, and to operate a videotape recorder, a projector, and a film camera.

Nervousness. Most performers are nervous before a show, and some are extremely nervous. This is because they want so strongly to do well. A performer must decide whether the nervousness is detrimental to the act or to the performer's health. The best performers, once they are secure on camera, experience a degree of abandon and self-control, then confidence, enjoyment, and finally euphoria. Lack of concentration, improper preparation, and insufficient desire to enchant or inform an audience are the main reasons for nervousness. Most good performers use nervousness to their advantage. It makes the adrenalin flow and gives energy to the performance.

Preparation. Most novice performers who claim they want to be on the air are inadequately prepared when the opportunity comes. Opportunities come when they are least expected. A smart performer will have a modest financial reserve before leaving for New York City or Hollywood. The reserve, used only for emergencies, prevents one from having to live on the streets.

Stamina. Performers, as a rule, are healthy despite very long, irregular hours and highly repetitive work. Performers cannot afford to be sickly; they must be dependable. Performers should practice good health habits to maintain physical endurance.

Time. There is never enough time to do the massive amount of work television demands. Performers quickly learn to do whatever they can as well as they can and move on to the next project without remorse.

Voice Quality. Performers tend to neglect their voices in favor of their appearance. Both are equally important. An arresting voice will get as much attention as a unique look.

BIBLIOGRAPHY

Barr, Tony. *Acting for the Camera.* Boston: Allyn and Bacon, Inc., 1982.

Hawes, William. *The Performer in Mass Media*. New York: Hastings House, 1978.

Hindman, James, Larry Kirkman, and Elizabeth Monk. *TV Acting, A Manual for Camera Performance*. New York: Hastings House, 1979.

Hyde, Stuart W. *Television and Radio Announcing*. 3d ed. Boston: Houghton Mifflin, 1979.

Keith, Michael C. *Broadcast Voice Performance*. Boston: Focal Press, 1989.

Trade publications serving performers include:

Billboard, *Broadcasting*, *Cash Box*, *The Hollywood Reporter*, *Variety*.

Glossary

Above-the-line costs Budget for the so-called creative elements in the production, such as writer, director, performer, designer; differentiated from below-the-line costs.

Action Cue from the director to the performers and technicians to begin the sequence.

Affiliate A television station that has contracted with one of the networks to receive a specified minimum of hours of that network's programming per week.

AFTRA American Federation of Television and Radio Artists; a performers' union.

AGVA American Guild of Variety Artists; a performers' union.

AM band Amplitude modulation. (See *modulation.*)

Analog Nondigital transmission by amplitude modulation (AM) or frequency modulation (FM).

Angle The camera's position when taking a shot. The terms *low-angle shot* and *high-angle shot* are used frequently.

ARB American Research Bureau, one of the major audience survey and ratings services.

Arbitron The arm of the ARB responsible for television viewer research and ratings.

Arc The camera's curving movement as it takes a shot.

Audience flow The carryover of the audience from one program to the next program or time period. Programmers try to maintain a high flow through similar programming.

Audimeter A.C. Nielsen electronic device to measure the number of television sets tuned in to a given program.

Band A specific segment of the spectrum allocated for the use of a designated service. For television the band is broken up by the FCC into channels, each of which can accommodate a signal of 6 MHz.

Back timing When determining how to fit a show into a required time length, the director starts at the end of the script and times backward, figuring out where in the script the action should be at specified times.

Barter A station provides commercial announcements in exchange for the carriage of a television program. The commercials may be on the program or put elsewhere in the station's schedule. This is in lieu of advertisers paying cash fees for the ads.

Below-the-line costs Budgeting the technical costs, such as crew, technicians, clerical personnel. Differentiated from above-the-line costs.

Binary system A numerical system using only two digits (0, 1) in combinations to represent any possible value.

Block programming In order to create effective audience flow, programmers sometimes put several similar programs together in a given daypart.

Blocking The director's moving of performers into the places required for the most effective visual pictures and character relationships.

Broadcast quality Industry standards establishing the expected transmission quality for commercial broadcasts.

Budget A written statement of anticipated income and expenditures for a future specified period, used as the basis for station and individual department financial operation.

Business Specific actions by the performer, frequently assigned by the director.

Camera chain The camera and its power source.

Carrier A wave used to carry signals from the station's antenna to home receivers.

CCD A charged coupler device; used as chips in cameras instead of tubes.

Channel A passage through which a message (or signal) is conducted.

Clearance An affiliated station clearing a network program for local broadcast. The number of clearances determines, along with ratings, whether a given network program will remain on the air.

Close-up Or close shot. Usually filling the screen with the subject. Extreme close-up usually results in just the character's face or other part of the body or object filling the screen.

Coop advertising The cost of the commercial is shared by the manufacturer and the distributor or seller. Sometimes used to refer to several advertisers sharing the cost of a program.

Counterprogramming Finding a program that will surpass or at least challenge the rating of a competing program in the same time slot.

CP Construction permit from the FCC to build a station. Only after equipment tests may the station receive permission for program tests and then a license.

CPM Cost per thousand. How much money it costs the advertiser to reach each 1,000 television households or viewers.

Cross A performer's movement from one place to another on the set.

Cue The signal to the performer to begin the action.

Cume How many viewers saw at least five minutes of a given program.

Cut In television, the instantaneous switching from the picture on one camera to that on another. In film, a command to stop the shooting.

CVC A compact videocassette used in ¼ inch videotape format.

Daypart The division of the broadcast day into categories of hours.

DBS Direct broadcast satellite. The obtaining of programs with a satellite receiving apparatus direct from the source rather than through a local station or cable system.

Demographics The analysis of the many variables making up an audience, such as age, gender, education, economic status, and other factors that help a station and advertiser determine the kind of programming and product or service most likely to appeal.

Digital Describes any equipment using a system of encoding a signal with binary impulses.

Dissolve As one picture disappears from the screen the next picture appears. Similar to the *cross-fade,* where the second picture comes onto the screen as the first one goes to black.

Dolly The moving platform on which the camera sits. As a verb, it means the moving of the camera, such as "dolly in" and "dolly out."

DTTR Digital television tape recording, as opposed to analog recording.

Dub A duplicate video or audio tape recording.

Earth station Sends (uplinks) programs to a communication satellite and/or receives programs (downlinks) relayed from a satellite.

EFP Electronic field production.

EFX Visual effects, sometimes referred to as special effects.

ENG Electronic news gathering outside the studio.

Establishing shot Setting the scene, providing basic information for the audience, usually through the opening shot.

EXT Refers to exterior setting in a film script.

Fade Fade-in is the appearance of the picture from a black screen. Fade-out is the reverse. Used to indicate the passage of time and to begin and end the show.

FCC Federal Communications Commission, established by the Communications Act of 1934 to regulate all domestic private sector (nonfederal government) communications.

Fidelity Faithful reproduction of sound or visual signals.

Floor manager In effect, an assistant director who is on the studio floor giving directions to the cast and crew, received by earphones from the director in the control room.

FM Frequency modulation. (See *modulation.*)

Frequency (1) The specific channel authorized for transmission. (2) How many times a viewer or television household is exposed to a given commercial.

Frequency response Pertains to the range of frequency capabilities of a given product (i.e., the height and depth of its parameters and level at which its quality is best).

From the top Usually during a rehearsal, to start from the beginning.

Gigahertz One billion cycles per second.

Ground station See *Earth station.*

HDTV High-definition television. An enhanced wide-band system producing a picture of 1,125 horizontal lines that offers superior detail over the present U.S. standard of 525 lines.

Hertz The unit measure for frequency (i.e., one Hertz is one cycle per second).

IATSE International Alliance of Theatrical Stage Employees and Motion Picture Operators of the United States and Canada.

IBEW International Brotherhood of Electrical Workers.

Idiot card Cue cards, with dialogue written in large letters, held up for performers who have not memorized the lines.

Input The signal from a given source, such as a microphone, tape, or any other equipment fed into the console or mixer.

Institutional advertising Selling the image of the company, rather than a specific product or service.

INT Refers to an interior setting in a film script.

Interface The matching of one machine to another so that they may function as a team.

Integrated circuit A miniaturized electronic chip with more than one capability.

Inventory The stock of time a station has available to sell to advertisers.

Key The combining of two different visual sources into one picture. (See *matte.*)

Limbo Where the performer stands out from the background through special effects or lighting.

Long shot A picture of the entire scene. Frequently used for an establishing shot.

MAC Multiplexed analog component; a system modified by 525-line, 60-field system-M signals.

Market research Determining the factors in the community that are most conducive to marketing a product or service. Conversely, determining factors that bode ill for the product or service.

Marketing Providing a product or service in exchange for something of value (i.e., cash, goods); includes buying, selling, research, investing, advertising, promoting. A station markets itself, specific businesses, or items, even television itself.

Matte Same as *key,* but adding color.

MegaHertz One million cycles per second.

MERPS Multiple event record and playback system.

M-11 M-signals developed by Matshusita; referred to as M-11 format using metal particle tape instead of oxides.

Modulation The act of superimposing a signal onto a carrier wave so that it may be transported from one place to another. AM *(amplitude modulation)* occurs when a signal modulates a carrier in such a way that the height (amplitude) of the carrier wave changes. FM *(frequency modulation)* occurs when a signal modulates a carrier wave by changing how frequently it repeats its wave form. PM *(pulse modulation)* uses only samples or sections of the signal, known as pulses, to modulate the carrier. There are several types of pulse modulation. Digital utilizes pulse code modulation (PCM).

MTS Multichannel television sound; used in processing audio stereo with a stereo synthesizer. Used by MTS television stations.

MX Mutually exclusive. In FCC use, when two or more bona fide applications are received for the same channel.

NAB National Association of Broadcasters. The lobbying organization for commercial stations.

NABET National Association of Broadcast Employees and Technicians.

NCTA National Cable Television Association. The lobbying organization for cable systems.

Nielsen, A.C. A leading audience research and ratings system. Uses meters in homes to monitor set use.

Noise Any unwanted video or audio distortion.

NTSC National Television Standards Committee.

O & O Owned and operated stations. Those belonging to a network or other group owner. In 1988 the limit for a single owner, with certain exceptions, was twelve AM, twelve FM, and twelve TV stations.

Organizational chart A diagram showing all the major offices/positions in a station and their hierarchical relationships to each other.

Pan The right or left movement of the camera to take a shot or when shooting.

PBS Public Broadcasting System. An association of public television stations for acquiring and distributing programs.

PCM A kind of pulse modulation that involves coding each pulse (see *modulation*).

Pilot The first program in a potential series. Reactions to the pilot usually determine whether the series will be bought and aired.

Postproduction When the editing and other technical requirements, such as dubbing and scoring, are done.

Preemption A station's cancelling a network program after clearing it; a network cancelling a program after offering it; a station's switching or cancelling of a discounted ad if a full paid commercial materializes for the time slot.

Preproduction Completion of all the requirements prior to shooting the program, such as finalization of the script, casting, set design, lighting, sound crew orientation, exterior locations, and other elements.

Prime time The television daypart from 7:00–11:00 P.M.

Promo An ad by a television station for one of its own programs.

PTAR Prime time access rule. With certain exceptions, limits stations to three hours of network programming during prime time. This limitation effectively results in prime time network programming running from 8:00–11:00 P.M. (7:00–10:00 P.M. in the Central and Mountain time zones).

Racking Changing the focus of a camera from the foreground to the background and vice versa.

Rating The percentage of all television sets tuned to a given program.

Remote A program being produced at a site away from the studio and carried live.

Roll cue The director's command to start the tape or film required in a program, including commercials.

Routine sheet A detailed outline of the different segments of a program, as differentiated from a complete script.

Rundown sheet Similar to the routine sheet, but sometimes not as detailed.

SAG Screen Actors Guild. A performers' union.

SAP Secondary audio program channel used for stereo television.

Satellite A communication satellite stationed 22,300 miles above the earth and able to relay radio and television signals to any number of earth stations within each satellite's coverage area. Each satellite has several transponders and directional antennae enabling simultaneous delivery of several signals from one satellite by means of spot beams aimed at particular regions.

Scale Minimum union wages for a performer.

SFX Sound effects.

Share The percentage of total television sets on at a given time tuned to a given program.

Signal A wave carrying intelligence.

Signal-to-noise ratio A statement of the relationship of wanted sound output to unwanted noise.

SMPTE Society of Motion Picture and Television Engineers.

SNG Satellite news gathering.

SOF Sound on film.

SOT Sound on tape.

Spot television The purchase of commercial time on a local station.

Stand by Director or floor manager cue to cast and crew to be ready to run the show; usually thirty seconds warning.

Station rep The national or local representative selling commercial time for the station.

Storyboard The script illustrated by a series of drawings, sometimes each drawing signifying successive shots.

Stripping Running the same program series in the same time slot every weekday.

Sweeps A quarterly ratings study of the entire country's television viewing by Arbitron and Nielsen. Networks gear up their best shows for presentation during the sweeps period.

Switcher The control room device that creates the various effects and moves from one visual to another, such as a dissolve or a cut between cameras.

Syndex The right of a station or cable system to purchase exclusive rights to a program or series for its area of service, including the exclusion of cable distant signals carrying that program.

Syndication The selling of programs to individual stations or markets rather than to a network.

Tally lights Small lights on the camera that are lit when that camera's shot is on the air.

TBC Time base corrector, which records on a videotape to provide each video frame with its own identification indicating the hour, minute, second, and frame number. Used in video postproduction.

TD Technical director. The person in charge of the operation of the control board/switcher.

Tracking When the camera moves with the performers in the same direction. Sometimes called *truck shot.*

Transponder That part of a communication satellite used to receive signals from an earth station and to beam them down to other earth stations. In broadcasting, used to relay broadcast programs.

VO Voice over; dialogue read by a performer over silent live action, film, or tape.

Wipe An emerging picture on one part of the screen, pushing the existing picture off in any direction or angle.

Wrap The reverse of *stand by.* A warning cue that the action will end in thirty seconds; sometimes called a *wind up* cue. The conclusion of shooting for the day.

Zoom The movement of the camera lens from a narrow-angle to a wide-angle shot and vice versa. Instead of the camera dollying in or out, the zoom moves the subject closer or farther away.

Index